INDIVIDUAL DEVELOPMENT FROM AN INTERACTIONAL PERSPECTIVE: A LONGITUDINAL STUDY

PATHS THROUGH LIFE

A series of volumes edited by David Magnusson

Volume 1
Individual Development from an Interactional Perspective: A Longitudinal Study

PATHS THROUGH LIFE
Volume 1

INDIVIDUAL DEVELOPMENT FROM AN INTERACTIONAL PERSPECTIVE: A LONGITUDINAL STUDY

DAVID MAGNUSSON
University of Stockholm

LEA LAWRENCE ERLBAUM ASSOCIATES, PUBLISHERS
1988 Hillsdale, New Jersey Hove and London

Lawrence Erlbaum Associates, Inc., Publishers
365 Broadway
Hillsdale, New Jersey 07642

Library of Congress Cataloging-in-Publication Data
Individual development from an interactional perspective.
(Paths through life ; v. 1)
Bibliography: p.
Includes index.
1. Developmental psychology—Longitudinal studies.
2. Nature and nurture—Longitudinal studies. 3. Developmental psychology—Research—Sweden. I. Magnusson, David. II. Series.
BF713.I52 1987 155 86-24016
ISBN 0-89859-707-2

Printed in the United States of America
10 9 8 7 6 5 4 3 2 1

CONTENTS

PREFACE

This book presents a longitudinal research project "Individual Development and Adjustment" (IDA), planned and implemented at the Department of Psychology at the University of Stockholm. The first phase of the project (1965–1975) has been described by Magnusson, Dunér, & Zetterblom (1975). The present volume, which is the first in a series from the project, concerns the theoretical background of the project, the planning and collecting of data during the second phase of the project when the participants had reached adulthood, and the presentation of some empirical, illustrative studies based on the collected data.

Several people have contributed to the success of this second phase and I would like to take this opportunity to thank them for their efforts. First I would like to thank Dr. Hans Linderoth and Dr. Per Olov Ganrot. Dr. Linderoth is director of the medical health care organization of the local school system in the community from which the three cohort studies come. Without his continuous, active support of the project, it could not have been carried through as successfully as it has. Dr. Ganrot is director of the regional hospital and made it possible for us to use the laboratories for medical examinations, and the testing and interview sessions that formed an important part of the follow-up studies.

The entire staff of IDA at that time—Anders Dunér, Gunnel Backenroth, Ola Andersson, Helen Jakobson and Håkan Stattin—participated in the planning and implementation of the comprehensive survey study. The tedious collecting of data from official records and the analysis of these data were led by Håkan Stattin (IDA, criminal records), Katarina Bremme (Karolinska hospital, obstetric records), Anne-Liis von Knorring (University of Umeå, psychiatric records), Hans Bergman (Karolinska Institute, alcohol records), and Tommy Andersson (IDA, alcohol records).

An important component of the follow-up of the main cohort at adulthood was individually performed medical examinations, tests and interviews. Professor Hans Bergman (Karolinska) and Professor Daisy Schalling (Department of Psychology, University of Stockholm) contributed actively to the planning and collecting of data in testing and interview sessions. Continuous cooperation with medical researchers was a prerequisite for the medical examinations. They were planned and carried out in cooperation with Professor Peter Eneroth (Karolinska Institute, hormone laboratory), Professor Margareta Blombäck (Karolinska Institute, blood analysis laboratory), Professor Lars Oreland (Departments of Pharmacology, Universities of Uppsala and Gothenburg) and Professor Lars A. Carlsson (Karolinska Institute, Gustav V's research laboratory). Professor Eneroth coordinates the medical part of the project with the assistance of Monica Lagerström (IDA).

All members of the project have contributed in one way or another to the production of this first book in the series. I particularly want to thank Håkan Stattin for his contribution to the presentation in Chapter 6 and Lars R. Bergman for his contribution to Chapter 8. Ola Andersson has taken responsibility for the administration of the data base and the computer work. Boel Bissmarck, Viera Dornic and Maj-Britt Palmers have worked hard to prepare the manuscript. Much of the editorial work has been taken care of by Bertil Törestad and Agneta Willans. Tensie Whelan checked the English and contributed with her skill to the editing of the manuscript. I thank them all, not only for their contributions to the production of this volume, but also for the personal qualities of competence, dedication to scientific work and generosity they have shown in the daily work of the project. The late Vernon L. Allen was a knowledgable, stimulating and generous discussion partner in the planning of the book.

An early version of the manuscript was read, in parts or the whole, by colleagues who generously shared their critical comments and suggestions; Particularly I want to thank Jack Block, Robert Cairns, Alex Kalverboer, Richard Jessor, Donald Peterson, Michael Rutter and Saul B. Sells. Their comments were highly valuable in the preparation of the final version of the manuscript.

The research presented in this volume was made possible by financial support from the Bank of Sweden Tercentenary Fund, the Swedish Council for Planning and Coordination of Research, and the Swedish Council for Social Research.

David Magnusson

FOREWORD

At the present stage of developmental study, the wisdom and insights of the investigator are as important as the procedures adopted. It is commonly assumed that data speak for themselves—especially if they are prompted by meta-analysis and multivariate statistics. That is an illusion. The story told by any longitudinal data set is necessarily filtered through several layers of method, statistical manipulation, and inference, then narrated and amplified by the reporter. So readers must trust that investigators are in tune with the meaning of their findings, not with the glitter of probability values. On this count, the present volume integrates the best of both worlds. David Magnusson is not only a rigorous scientist, he has a keen grasp of the human condition won through the several years he spent teaching in school classrooms and working as a school psychologist. No 'invisible investigator', Magnusson is sensitive to the uniqueness and meaning of the lives of individual children. This personal concern reflects the spirit of the project, and careful science guides its conclusions.

Recently my wife and I visited David Magnusson and his research group in Stockholm. It was about as we had envisioned; a closely-knit team of scientists who dealt creatively with developmental concepts and attended carefully to empirical details. Possibly because of his prior work on test theory, Magnusson harbors a healthy skepticism about the limits of traditional methods of developmental measurement. His concern is simply that greater statistical sophistication is called for, not less. In the present volume, for example, the clustering procedures and configural frequency analyses have melded advanced statistical procedures with a broad concern for a person-oriented perspective.

This work is one of the few modern longitudinal projects to have been based on an explicit developmental formulation; hence the theoretical framework requires careful explication. Magnusson describes the concepts with economy and clarity. As he points out, it has become common to decry static, mechanistic learning theory and to embrace organismic, transactional models of human development. So far, so good. However, gaps often arise between organismic theoretical pleas that stimulate the research and the mechanistic procedures that are employed in its execution. Here the Swe-town longitudinal study is different. As the first part of this book argues, it is possible to align systematic developmental theory with appropriate developmental methods. Possible, yes; usual, no. Historically, the problem has been that the methods and constructs available for longitudinal study have homogenized true developmental changes. By employing measures that are appropriate for different age levels, Magnusson has been able to track continuities and determine where novel behavior patterns emerge.

The need to employ multiple levels of analysis is recognized throughout this longitudinal series. Each level first addressed the theoretical issue for which it was appropriate, then the levels have been integrated to answer specific questions of research interest. Moreover, psychobiological events can play an integral, dynamic role in social adaptation at any developmental stage. For example, we find how the very onset of menarche can affect the pubertal girl's membership in social cliques, and how these social associations can shape social values, sexual behaviors, and school performance. Similarly, we learn why correlations between adrenaline and aggression tell only part of the story of eventual social adaptation and the occurrence of criminal behavior.

David Magnusson and his colleagues anticipated many of the subtle problems of longitudinal study when the foundations for the investigation were laid in the early 1960s. Some of the measurement innovations they introduced were novel at the time, and they remain fresh today. This includes, for example, the "shrinking classroom" method for the assessment of social structure, the collaboration of the Karolinska Institute in the biophysical assessments, and the involvement of an entire community to recruit, participate, and retain subjects. Equally important are the theoretical insights embedded in the structure of the research design. As Magnusson observes, ontogeny necessarily entails the emergence of novel characteristics and behavioral patterns, and it would be folly to expect a given measure to assess the same characteristics at widely disparate ages. Significant aspects of adult years may be seen only darkly by investigators who peer exclusively at adaptations in early life.

Many of us in America believe that longitudinal work in Scandinavia must be an easier pursuit than comparable study at home. On some levels, the belief is justified. Subjects in Sweden are readily traced year-to-year, and national records on arrests, mental health, and marriage and divorce are more accessible than in most states. But it is also the case that certain perils—including socially organized disruption and challenge—are just as great, if not greater, in Sweden than in the United States. Hence the care taken to inform subjects and involve the community proved to be good science and to make good sense.

In brief, this 20-year study of more than 2,000 children, their parents, and their teachers is a model longitudinal investigation. On the standard criteria—namely, breadth of sampling, rigor in execution, and care in followup and analysis—the work is destined to be ranked with the major studies of this century at Stanford, Berkeley, Oakland, Cambridge-Somerville, London, and Fels. It is also one of the few long-term longitudinal studies where the investigator who conceived of the study also organized and interpreted the results. Rarely does a researcher in any scientific field devote over 20 years to a single empirical investigation as David Magnusson has done. Such commitment is called for if developmental science is to have any hope of advancing beyond the commonplace.

Robert B. Cairns
Chapel Hill,
North Carolina

PROLOGUE

In 1964, at the University of Stockholm's Department of Psychology, a longitudinal research project entitled *Individual Development and Adjustment*[1] (IDA) began and continues today.

The purpose of the project is to investigate how person and environmental factors operate and influence the course of individual development from childhood to adulthood. Within this general frame of reference, subprojects have been directed to the study of the developmental background of maladjustment at an adult age, that is, alcohol problems, criminal behavior, and mental problems. These issues cannot be effectively studied without a solid understanding of normal developmental processes. Thus, the basis for the study of these aspects of the developmental process resulting in adult functioning is an understanding of normal development as a process of temporary and long-term adaptation to situational and environmental conditions. Special attention has been paid to the developmental role of individual educational and vocational careers.

At the start of the project, boys and girls enrolled in the third, sixth, and eighth grades of a community school system in Sweden were studied, three cohorts in total. The majority of the subjects were 10, 13, and 15 years of age for each respective cohort. Each age group contained about 1,000 to 1,200 boys and girls with a fairly equal distribution between the sexes.

For this project, it was crucial that the cohorts were representative of the

[1]This is the original title. For some time the title *Individual Development and Environment* was used. From this point on, the current title *Individual Development and Adjustment* will be used as it better reflects the primary direction of the project.

population. Because there are no private schools in the community studied, and virtually none in Sweden, the cohorts represent unselected groups of youngsters. The cohorts are described in more detail in Chapter 4.

The oldest age group, those who were 15 years of age at the start of the project, were only studied at the first data collection in 1965. This group is called the *pilot* group. The second cohort was followed from the age of 13 at the first data collection to the age of 23, with successive data collections during the time at school and one main data collection at adult age. This group is called the *initial* longitudinal group. Besides yielding data that are of interest in themselves for the elucidation of the developmental process, these data collections helped to plan for research on what has been designated as the *main* group, those who were 10 years of age at the start. This group was followed continuously during the school years, and a comprehensive data collection was implemented when the subjects were 26 to 28 years of age.

A PERSONAL BACKGROUND

Just as behavior of any individual depends on personal experience, so does the direction and frame of psychological research depend on the prior experience of its conductor.

My interest in individual development as an object of scientific inquiry, and consequently how individual development was approached in the longitudinal project, has three main roots: experience as an elementary school teacher and school psychologist, cross-sectional studies on adjustment problems among children, and both basic and applied psychometric personality research.

Before going to the university, I worked for six years as an elementary school teacher in a small town in southern Sweden. During my undergraduate studies I worked for two and a half years as a school teacher and as a school psychologist in a suburb of Stockholm, one part of which was an old slum. The years that I spent with children aged 9 to 15 provided me with an experience that influenced the planning and implementation of IDA. As a teacher of "normal" children, I learned how widely children differ in almost all aspects of individual functioning. Among other things, I became aware of the low level of abstract thinking at which many children function. Happily, I also learned that even children with low intellectual resources can be helped to learn a great deal if they are treated as individuals and if learning takes place in a planned fashion and at a pace that is adjusted to individual capacity. As a teacher of retarded teenagers, I found that children, who in the frame of reference of a "normal" class look like a grey homogeneous mass of low potential, have strong and important differences among them which make them as unique as "normal children." As a school psychologist, I had many opportunities to observe the large range of conditions under which boys and girls are brought up. These experiences taught me much about the role of environmental conditions in understanding why some boys and girls behaved as they did. At the same time, I found how little help was offered by the textbooks in psychology for understanding the processes underlying various forms of social maladjustment.

During these years, I also had some specific experiences that influenced my view of research on individual development. One of my pupils underwent a remarkable change of personality following an accident in which a portion of his frontal lobes was damaged. Before the accident he was a careful and conscientious pupil of average intelligence. After the accident he neglected his homework, started lying, became generally careless about his work, and his achievement at school declined noticeably. In another class I had a 14-year-old boy who was described by his earlier teachers as extremely lazy. No motivation theory could help me motivate him for school work, though he seemed to like school. When I sent him in for a medical examination, it was found that his thyroid gland was malfunctioning. These and similar experiences taught me the importance of biological factors in individual functioning.

Experience gained from cross-sectional studies on adjustment among school children stimulated my interest in this area of research. One study was concerned with self-perception and achievement at school (Magnusson, 1960). In this investigation, a method for the study of social neglect and rejection (later applied in IDA) was worked out. A second study was designed to determine why some individuals at a low level of tested intelligence could meet school requirements with respect to achievement and be able to stay in a "normal" class, while others at the same level of intelligence had to be taught in special classes (Magnusson, 1963). The study was performed using a matched group design in which intelligence, age, and social background were controlled. Among other things, this study showed that those in the "clinical group," that is, boys and girls who were referred to special classes, had significantly more errors on both the Bender and the Benton psychomotor tests. This study indicated that cognitive-perceptual processes might be involved in the process underlying malfunctioning with respect to underachievement and conduct. A third pilot study was concerned with the relationship between relative academic achievement, on one hand, self-perception, neglect and rejection by peers and various aspects of misconduct in the classroom, as rated by teachers, on the other hand. The main finding was that misconduct as rated by teachers showed a rather strong, systematic relationship to under-achievement when defined as the difference between actual achievement at school and achievement predicted from general intelligence. Experiences with this rating procedure played an important role in planning the procedures for data collection in IDA.

When I tried to interpret the results of these and other studies on children of various ages, in which I applied the psychometric tools available to me, I became more and more convinced that in order to understand the process underlying the relationships found in the cross-sectional data, it would be necessary to follow the same individuals over a longer period of time, that is, to conduct a longitudinal research project.

The third experience that influenced the planning and implementation of IDA came from psychometric work and application of classical test theory during applied clinical activity and selection (Magnusson, 1967). The basic idea, that it was possible to understand individual mental functioning and behavior solely by reference to presumably persistent personality dispositions or traits, became more and more unrealistic, both from the viewpoint of everyday experience and in terms of the theoretical analysis of psychological processes.

With some of my students, I planned a series of studies concerned with the extent to which individual behavior is differentially dependent on situational conditions. These studies led to a project on person-situation and person-environment interactions that is still active (e.g., Ekehammar, 1974; Endler & Magnusson, 1976a, 1977b; Heffler & Magnusson, 1979; Magnusson, Gerzén & Nyman, 1968; Magnusson & Heffler, 1969; Magnusson, 1975, 1976, 1980; Magnusson & Allen, 1983a, b; Magnusson & Endler, 1977a; 1977b; Magnusson & Öhman, 1986; Magnusson & Stattin, 1981a; Öhman & Magnusson, 1987). The experimental studies in quasi-natural situations with systematic variations of conditions convincingly demonstrated that it is not enough to know the individual's general trait dispositions if we want to understand or explain his/her behavior in a specific situation. For some purposes, it is also necessary to know something about the character of the specific situation in which the individual is functioning. This implies, among other things, that data from traditional non-situation-specific inventories, aimed at measuring individual dispositions at a rather high level of generalization of individual functioning, must be complemented with situation-specific data, that is, data referring to individual functioning in specific situations or specific types of situations under conditions that can be understood and controlled in a planned and known way.

One of the implications of the theoretical formulations about the role of situational conditions for current behavior as well as for development is that environmental conditions have to be considered in planning and interpreting empirical research on individual functioning. However, successful research on the role of situational-environmental conditions in the person-situation interaction process presupposes that the environmental conditions are organized in meaningful structures that can be used systematically to plan and interpret empirical research.

The central problem, then, is a description of the environment and the situations in which an individual appears, in terms of the individual's perceptions and interpretations of the environment. In order to obtain such descriptions a suitable methodology is needed. This became a central goal behind our project work on person-situation interactions during the 1970s and the beginning of the 1980s (e.g. Ekehammar & Magnusson, 1973; Magnusson, 1971, 1974, 1981a, b;

1984a, b; Magnusson & Ekehammar, 1973; Magnusson & Olah, 1981; Magnusson & Stattin, 1978, 1981b; Magnusson & Törestad, 1982; Nystedt, 1981; Stattin, 1983; Stattin & Magnusson, 1981; Törestad, Magnusson, & Olah, 1984). Most of this research has been focused on stress and anxiety, although recent research has also dealt with anger.

In order to put our own research into an interactional perspective, four international symposia have been organized and reported. They have been primarily concerned with general models concerning the individual in the person-situation process (Magnusson & Endler, 1977a), models focused primarily on the situation (Magnusson, 1981a), models of development as related to person-environment interactions (Magnusson & Allen, 1983a), and models for the study of psychopathology as related to person-environment interactions (Magnusson & Öhman, 1987).

Without the experiences I have just discussed, IDA would not have been planned and carried through in the way it was. They influenced the choice of aspects of individual functioning to be studied, the kind of instruments and procedures used for data collection, and the procedures introduced for continuous contact with and information to parents, subjects, teachers, and school authorities.

An important factor in the start of IDA was the collaboration with an early friend of mine, Anders Dunér, who worked in the same department. We had studied at the same Elementary School Teachers' College and shared common experiences as elementary school teachers. Anders Dunér participated actively in the planning of the project and served for several years as the administrative project leader. His own particular interest has been directed to the developmental process of educational and vocational careers, an area in which he is still active (Dunér, 1978a, b).

PREREQUISITES FOR LONGITUDINAL RESEARCH

It is relatively simple to collect data by employing "quick and easy" procedures (tests and questionnaires), and, given access to computers, it is easy to calculate data outcomes. However, effective research that can provide real contributions to knowledge about the phenomena under consideration requires careful planning and analysis. The essential conditions for an effective elucidation of the developmental process by longitudinal research can be summarized as follows:

1. A *common theoretical frame of reference, within which the empirical research can be planned, performed and interpreted.* The theoretical perspective that has guided the planning, implementation and interpretation of the em-

pirical component of IDA can be called an interactional paradigm. It is briefly described in Chapter 1.

2. *Careful analyses of the characteristic features of the developmental process at various stages, built on available knowledge and relevant theories.* Such analyses are necessary in order to determine the relevant basis for the choice of aspects of the developmental process to cover at each stage. The aspects of an interactional perspective, which were of special importance for the planning of the longitudinal project, are considered in Chapter 2.

3. *Careful analyses of methodological issues.* These are of two types. First are the problems of choice and design of methods for observation and data collection. The choice of method for the elucidation of a certain problem is dependent on the type of variable covered, the level of complexity of the structures and processes under consideration, the age level of the subjects and numerous other factors. Second, crucial methodological issues arise in developmental research concerning models and methods for proper and effective analysis of data. The appropriate method for treatment of data depends on the level of analysis and generalization, on the type of data available, and on other characteristics and inferences to be drawn from the data. The methodological considerations for an interactional perspective concerning the study of individual development in a longitudinal perspective are discussed in Chapter 3.

4. *An effective research strategy with regard to the appropriate age level for observations of various factors under consideration.* Research strategy considerations for our longitudinal project are discussed in Chapter 4.

THE PURPOSE OF THE BOOK

In 1975, ten years after the start of the project, two of my coworkers and I reported in detail the planning and conduct of the investigations during the first phase of the project (Magnusson, Dunér & Zetterblom, 1975). The instruments and administrative procedures used for data collection, the psychometric characteristics of the data, the cohorts and their general characteristics, and so forth, were also reported. The report included data analyses on the following topics: norms and norm conflicts among teenagers; delinquent behavior in boys; social adjustment and intrinsic adjustment among adolescent girls; choice and lines of academic study and occupation; creativity and adjustment; social differentiation; biological factors and adjustment and; social relations and adjustment.

At the time of preparation of the first book, most of the youngsters in the main group were still in school or had just finished school. More than ten years have elapsed since that time. The males and the females of both the initial group and the main group have now reached adult life and have become established in working life.

PLAN OF THE BOOK

Part I of the book contains four chapters concerned with the theoretical background of the planning, implementation and interpretation of the project as well as its general design.

Chapter 1 presents the general theoretical framework for the project, which is based on an interactional perspective. It is an integration of three traditional metatheoretical approaches to the explanation of individual functioning—mental, biological, and environmental. The interactional perspective takes its point of departure from three basic propositions: (a) The individual functions and develops as a total integrated organism. Development does not take place in single factors or variables per se, in isolation from the totality. (b) The individual functions and develops in an ongoing, dynamic, and reciprocal process of interaction with his/her environment. (c) The character of the person-environment interaction process depends on the character of the ongoing, dynamic and reciprocal process of interaction among the individual's psychological and biological subsystems. The major aspects of an interactional perspective as a framework for theorizing and empirical research on individual functioning, both in a current and a developmental perspective, are discussed.

With an emphasis on points of particular interest for the planning, implementation and interpretation of the longitudinal project, more detailed theoretical and conceptual considerations of the general view are discussed in Chapter 2. These include: the study of persons vs. the study of variables as a direction for developmental research; the role of theory vs. observation in developmental research; prediction vs. explanation as the ultimate goal of psychological research; stability and change in development; the debate about the possible existence of "traits" in the interactional perspective of development; maladaptation in an interactional perspective and the importance of protective factors in social adjustment processes and; a distinction between extrinsic and intrinsic adjustment.

In Chapter 3, the general implications of the interactional perspective for the methodology and research strategy of the project are discussed. With reference to the discussion in Chapter 1, the importance of using data appropriate to the level of the processes under consideration is emphasized. Most of Chapter 3 is devoted to a discussion of methodological problems connected with the investigation of development in terms of single variables or factors. A central problem is the measurement of change. The use of linear regression models for the study of single variables is discussed against the background of the existence of interindividual differences in growth rate and of interactions among subvariables. With reference to the theoretical formulations concerning a holistic person approach in Chapter 1, the final section of the chapter is devoted to a discussion of appropriate methods for the study of development in terms of patterns.

The general characteristics of this theoretical framework have been the same from the beginning. This can be seen when one looks at the actual design of the project, the research strategy that has been followed, the choice of aspects of individual functioning that have been covered, the methods that have been used for data collections, and other aspects of the research. In general, the research strategy has remained constant, though the thinking has been refined within the original framework. We have learned from our experiences and analyses, from empirical research presented by other researchers and, above all, from the ongoing theoretical development in the field in which we, along with many other scholars, have been actively engaged. These factors require that the theoretical and methodological assumptions that have guided the work be presented explicitly and in a coherent way as a frame of reference for the presentation of studies directed to specific problems within the larger perspective of the total project.

The conduct of a longitudinal project such as IDA implies a series of important research strategy, methodological and administrative decisions. These concern the kind and size of sample(s) to be studied, aspects of individual functioning at various ages, choice of instruments to be used for data collections, administrative procedures for data collections, information for those involved (subjects, parents, teachers, school authorities), and ethical problems. The resolution of these problems is decisive for the outcome of the project.

As a background to a presentation of how these problems were solved in practice, Chapter 4 summarizes the main implications of the theoretical, conceptual and methodological considerations discussed in more general terms in Chapters 2 and 3 for the design and administration of IDA. The following *research strategy* implications are emphasized: (a) in order to understand important aspects of the process underlying individual development, we have to follow individuals across time, that is, conduct longitudinal research; (b) analysis of cohort effects should be possible and; (c) the sample chosen for the study should be representative with respect to basic characteristics.

With respect to *content* to be studied, the following implications are emphasized: (a) a broad spectrum of aspects of individual functioning (in psychological and biological respects) *and* of the environment should be covered; (b) the variables under consideration for each stage of development should allow for the specific character of that stage and; (c) the observation methods should allow for the level of the processes under consideration.

The rest of Chapter 4 discusses the major features of the planning and implementation of the project: the general design with three cohorts, the characteristics of the subjects, the data collections, ethical problems, and scientific cooperation. Most of the chapter discusses the procedures for data collection, the types of data obtained, and the variables covered by data. Because the data collections at school were reported in the earlier volume on the project (Magnusson et al., 1975), the emphasis in this presentation is on the collection of data from the subjects themselves in the follow-up at adult age, and the collection of data from official records.

Part II contains four chapters devoted to the presentation of empirical investigations performed in the project. Chapter 5 presents some general aspects of the life situation for males and females in the main cohort at adulthood. Data are presented that show the situation with respect to family characteristics, education, work, the participants' sense of control of their own lives, their leisure time activities and their social networks. Most of the chapter is devoted to the presentation of data from official records on criminal offenses, psychiatric care and alcohol abuse. Data for incidence and prevalence are given and results from some longitudinal analyses are presented. Among other things, the interdependence of the three aspects of maladjustment is investigated as a background for further analyses on the developmental history of adult maladjustment.

A main feature of the empirical analyses presented is that data were available for a large, representative group, with little or no dropout. The sample of studies was chosen to elucidate central issues in the project within the interactional frame of reference. The empirical studies demonstrate the way data have been used at various levels of generality, in a combination of cross-sectional and longitudinal studies, to fulfill the purposes of the project.

In the course of development, the individual interacts with an environment that is constantly changing at all levels of complexity. The individual partly contributes to the changes by direct action, by seeking new situations to enter, by taking on new roles in professional work, marriage and other aspects of life development, and, not least, by the role transitions that take place as an effect of the individual's biological maturation.

When Bronfenbrenner and Crouter (1983) discussed models for the role of environmental factors in the development process, they pointed to the emergence of dynamic models for the processes in which such factors operate, but concluded that the empirical research presented thus far had significant limitations. They presented two explanations for this state of affairs. They noted first that most research is cross-sectional, and that even if causal links are established, the second order effects may be purely situational, without a significant developmental effect on patterns of perception or behavior. Second, they suggested that research on the role of environmental factors for developmental processes should be directed to the study of *ecological transitions*. "An ecological transition takes place whenever, during the life course, a person undergoes a change in role either within the same or in a different setting" (Bronfenbrenner, 1979a).

The study presented in Chapter 6 is an example of the kind of research that Bronfenbrenner and Crouter considered fruitful. The central issue is the role of individual differences in biological age for the social development process in girls. This issue is investigated empirically in a current as well as in a longitudinal perspective. The age of the peers with whom a girl interacts during adolescence and their reactions to various types of norm violations are discussed as integral factors for the social adjustment process during adolescence. The chapter also reports the results of analyses of the long-term consequences for various aspects of adult life (education, marriage, and the number of children) of individual differences in biological maturation from a longitudinal perspective. The results empirically demonstrate a central aspect of the interactional perspective, namely the dynamic and reciprocal character of the person-environment interaction process.

The study reported in Chapter 7 illustrates several substantive and methodological issues. It focuses on conduct and physiological reactions at an early age (13 years of age) as important elements in the developmental background of adult social maladjustment. The first part of the chapter is devoted to an analysis of the longitudinal relationship between conduct as displayed by aggressiveness and motor restlessness (as an aspect of the syndrome of hyperactivity), and physiological processes as measured by adrenaline excretion, on the one hand, and adult criminality on the other. The correlation found between physiology and conduct at an early age and adult criminality directs the interest to more detailed cross-sectional analyses of the relationships among various aspects of cognitive functioning, physiological reactions, and conduct under standardized controlled situational conditions at an early age, based on situation-specific cognitive and biological data. The results of these analyses are reported in the second part of the chapter. The substantive and methodological implications of the results are discussed.

As an illustration of one of the main themes of the theoretical framework, longitudinal analyses of development in terms of social adjustment patterns, are presented in Chapter 8. The pattern analyses were performed for males based on critical configurations of data for main indicators of extrinsic adjustment problems at the age of 13 and data from official records for criminal offenses, psychiatric care, and alcohol abuse. Critical configurations of severe adjustment problems at the age of 13 as related to adult maladjustment are distinguished. The question of whether the existence of a small group of subjects with multiple severe adjustment problems can explain the strong correlation among maladjustment indicators (at an early age and at adulthood) is raised, and some tentative answers are provided. The results of the pattern analyses are briefly discussed, and further lines of research are suggested.

The book concludes with an Epilogue in which the nature of the book as the first in a series of volumes from the longitudinal project is reiterated and the main objectives for further research in the project are emphasized.

Part I

THEORETICAL BACKGROUND AND THE IMPLEMENTATION OF THE PROJECT

The theoretical, interactional background is presented and its theoretical and methodological implications for developmental research are discussed. Against this background, the planning and implementation of the project with respect to methods and procedures for data collection are presented.

Chapter 1

A THEORETICAL FRAMEWORK: THE INTERACTIONAL PERSPECTIVE

INTRODUCTION

As a scientific discipline, psychology deals with problems of crucial importance and interest for individuals, groups, and societies. Solid knowledge about psychological phenomena could contribute to the formation of childbearing conditions, educational systems, working conditions, and other major aspects of human life that better suit the needs and potentials of individuals and groups. Unfortunately, tangible contributions in this respect have been surprisingly few. An essential and challenging task for psychology is to analyze the reasons for this state of affairs. Such an analysis ideally could contribute to better performance in the future.

More and more researchers agree that one of the main impediments to real progress in psychology is the fragmentation of the field. It has been compartmentalized into subareas, each with its own concepts, methodologies and methods, and research strategies (Eysenck, 1983; Sameroff, 1982; Staats, 1981; Toulmin, 1981). One implication of this fragmentation is that empirical research in different areas is not planned, implemented, or interpreted in one general theoretical frame of reference. We lack both the basic conditions for the accumulation of knowledge that are characteristic of progress in a maturing science and a common language for the exchange of ideas. As a result, much psychological theorizing goes in circles instead of moving forward (Meehl, 1978).

Sometimes researchers describe the fragmentation of theory and empirical research in psychology as the result of specialization and regard it as a natural process in the development of a scientific discipline. What appears to be specialization, however, can have other explanations. In his evaluation of the status of the behavioral sciences, Toulmin (1981) discussed the problem of fragmentation and concluded:

> . . . while subdivisions of the physical and biological sciences into largely independent subdisciplines rests on a genuinely functional differentiation between their respective problems and issues, the fragmentation of the behavioral sciences rests—too often—on nothing more than sectarial rivalry and incomprehension. As a result, what has grown up in the behavioral sciences is less a rational and functional division of labor than a state of bureaucratic warfare! Even—given the charismatic character of the leading figures involved—a condition of sectarial hostility (pp. 267–268).

As far as this description is valid, it further emphasizes the need for a common perspective or frame of reference for psychological theory and empirical research.

The focus of the longitudinal project and the theme of this book are individual development. A common framework for developmental research should include a model of the human being that can be used as a framework for psychological theory and empirical research as a whole (Cairns, 1979a). One of the main points of departure for such a common framework is the definition of psychology, its goal and the delineation of the phenomena that are of central interest for psychological theory and empirical research.

The Goal of Psychological Theory and Research

Definitions of specific disciplines and delineations of their boundaries in relation to other disciplines are more or less arbitrary. The boundaries are generally neither self-evident nor obvious, and they change as a result of scientific progress. A good example is the development of physics and chemistry, where a very clear and strict distinction between the two has dissolved as a result of scientific progress in both fields.

At the same time, the central phenomena with which a certain discipline is concerned at its present state of development must be made clear. This clarification is essential, since the formulation of the goal of a scientific discipline has important consequences not only for theory but also for the way in which empirical research is planned, implemented, and interpreted. Making the goal explicit helps the researcher to see these consequences. The strong influence on the direction of development of a scientific field is illustrated by what happened to psychology after Watson (1913) formulated the goal of psychology as to "control and predict behavior."

The point of departure for a brief discussion of the theoretical framework for the present longitudinal project is our definition of the goal of psychology, a goal that connotes a strong heuristic value: *to understand and explain why individuals think, feel, act and react as they do in real life situations.* One important implication of this formulation is that it directs the main emphasis toward understanding and explaining the *lawfulness* of the way individuals function in real life (in the Wundt-James tradition) rather than toward *predicting* and *controlling* behavior (as in the Watson tradition).

The use of the goal outlined above necessitates a distinction between theories and models which analyze and explain individual functioning in a current perspective and those which employ a developmental perspective. Models that use the current perspective analyze and explain the way an individual thinks, feels, acts and reacts in terms of his/her current psychological and biological dispositions, independently of the developmental process that might have led to the present state of affairs. Developmental models analyze and explain current functioning in terms of the individual's developmental history.

Three Metatheoretical Approaches to Research on Human Functioning

The general theoretical perspective for the planning, implementation, and interpretation of the longitudinal project is interactional. As a background to a description of the main aspects of an interactional view, three main approaches at a metatheoretical level, to why individuals think, feel, react and act as they do will be sketched (cf. Magnusson, 1985a).

The first approach is *mentalistic;* it emphasizes mental factors. According to this approach, the main explanation for an individual's way of thinking, feeling, acting, and reacting is to be found in the functioning of the mind and can be discussed and explained in terms of intrapsychic processes of perceptions, thoughts, values, goals, plans, and conflicts. To this approach belongs the main stream of research on cognitive processes, information processing, decision making, and learning as well as personality theories such as most trait theories and psychodynamic models for individuals' current functioning. In developmental research this is reflected in the interest in the stability and change of personal characteristics such as cognition, intelligence, ego control, attachment, ability to delay gratification, and so forth. According to this approach, it is enough to investigate mental factors in order to understand why an individual thinks, feels, acts, and reacts as he does.

The second general approach identifies *biological* factors as having a primary influence on human behavior.[1] According to this reductionistic approach, indi-

---

[1]"Biological" is often interpreted to mean "inherited" in a sense that there is a significant correlation between generations with respect to the characteristic under consideration. However, it should be made clear that differences in specific aspects of individual psychological functioning that may be related to or dependent on biological factors do not necessarily imply that these differences are inherited in that sense. Assume, for example, that there was a perfect correlation between a

viduals' way of thinking, feeling, acting, and reacting basically can be traced back to biological factors and individual differences explained in those terms.

When current biological models are applied, an individual's thoughts, feelings, reactions, and actions are assumed to be determined presumably by his/her biological equipment and its way of functioning. Primary determining factors are assumed to be found in the physiological system, in the brain and the autonomic nervous system. An important aspect of the traditional application of this approach is the assumption that there is a unidirectional causality between biological factors on the one hand and mental factors and conduct on the other.

In biological developmental models, the major determining biological factors are genetic and maturational. For individual development, this view implies that individual differences in the course of development are caused by genes, with little emphasis on environmental factors. Cairns (1979a) described the extreme expression of this view in terms of the organism as a "gene machine" (p. 221). Hunt (1962) discussed and criticized this general view, which he summarized in terms of the concepts of "predetermined development" and "fixed intelligence", in his classic book *Intelligence and Experience*, and Scarr (1981) critically assessed it from the point of view of behavioral genetics.

The third approach locates the main causal factors for individual functioning in the *environment*. This general approach is reflected in theories and models of human behavior at all levels of generality for environmental factors: macrosocial theories, theories about the role of the "sick family" (Laing & Esterson, 1964) and S-R models for very specific aspects of behavior.

Experimental psychology in the S-R tradition is a good example of a current environmental model. Laws and principles for individual functioning are studied in terms of reactions to variation in the intensity of one or more aspects of the physical environment. Frequently, these studies are conducted in the laboratory under standardized conditions.

In developmental psychology, there are various environmental streams with different sources. One is the behavioristic model, which emphasizes the enormously strong role of the environment in individual development; Watson (1930) expressed this view when he wrote: "Give me a dozen healthy infants, well-formed, and my own specified world to bring them up in and I'll guarantee

---

certain substance in the physiological brain processes and musical ability. It would not necessarily follow that musical ability is inherited in the sense that there is a significant correlation between generations for musicability. For instance, the distinctive brain processes could have emerged during embryogenesis because of constitutional factors; alternatively, extended practice or experience could have provided feedback to increase the substance concentration. In either case, a developmental analysis is required in order to determine how constitutional, biochemical, genetic and experiential factors in the person are interwoven at any age.

to take any one at random and train him to become any type of specialist I might select—doctor, lawyer, artist, merchant-chief, and yes, even beggerman and thief, regardless of his talents, tensions, tendencies, abilities, vocations, and race of his ancestors" (p. 104). Another approach is rooted in sociology and is reflected in the vast amount of research on children growing up in environments that differ with respect to general, geographical or social characteristics such as social class, rural vs. urban cultures, or more specific characteristics such as one vs. two parent families, home vs. day care, mother's employment status, how many hours the father spends on child care and household tasks etc., in other words, what Bronfenbrenner and Crouter (1983) called the "new demography."

Like the biological model of a human being, one essential characteristic of the environmental model is the assumption of unidirectional causality, in this case between environmental factors and individual behavior. This characteristic is reflected in the planning and interpretation of research as well as in the interpretation of the many correlations presented between sociological characteristics of the environment and individual factors, particularly correlations concerned with various forms of extrinsic and intrinsic adjustment. For example, most textbooks on the developmental background of criminality now refer to environmental upbringing conditions as a main causal factor.

The distinction among the three approaches is not only a theoretical issue. It has important implications for planning and performing research on essential issues, as well as consequences for psychological application and political and administrative decisions and actions.

A good illustration of this point is the current discussion about the appropriate treatment of mental problems and mental illness. Each approach finds different consequences for cures and recommends different measures to forestall the development of such maladaptations. The mentalistic approach sees the main cause of an individual's suffering, from depression or schizophrenia, for example, as the malfunctioning of thought-processes: the natural treatment is assumed to be psychological therapy. According to the biological approach, an individual's thoughts, emotions, and actions can be influenced by changing his/her biological processes, and the appropriate therapy is therefore psychopharmacological treatment. The environmental approach holds that the genesis and development of various aspects of mental illness can be influenced and prevented by changing the environment through the removal, addition or improvement of external conditions. Another example can be found in the area of delinquency and the appropriate measures for prevention. Rutter and Giller (1983) argued that changes in the environment (the school, the area, the community, the physical environment) would be the most effective prevention. Farrington (1985) believed on the other hand, that it was as plausible to locate the causes of delinquency in the individual, a view that leads to other implications in the discussion about appropriate prevention measures.

THE INTERACTIONAL PERSPECTIVE

In psychological theory and empirical research, one basic prerequisite for overcoming fragmentation is a common theoretical frame of reference for planning, implementing, and interpreting empirical research. This should integrate, at a metatheoretical level, the three approaches presented briefly in the preceding section. A model that views the person as a psychological and biological being in constant interaction with his/her environment offers such a perspective (see Russell, 1970). Such an interactional perspective formed the frame of reference for the planning of the longitudinal project and has guided the research process from the beginning.

As a background to their presentation of models about heredity and environment in the development of temperament, Buss and Plomin (1984) discussed "interactionism" with reference to the following quotation from Thomas, Birch, Chess, and Robbins (1961): "Behavioral phenomena are considered to be the expression of a continuous organism-environment interaction from the very first manifestations in the life of the individual" (Thomas, Birch, Chess, & Robbins, 1961, p. 723). Buss and Plomin commented upon this formulation in the following way: "Surely, no one will disagree with this truism, but it is not very informative" (p. 28–29). This way of discussing an interactional position using single statements at a very general level and the resulting negative attitude are not uncommon. There are at least two possible explanations. First, many of those who take a critical position are obviously not very well informed; the way they discuss the interactional perspective very often reveals a superficial analysis of the essential issues and little knowledge about the relevant literature. Second, when it comes to synchronizing theoretical ideas, methods, analyses, and interpretations of empirical research into an interactional perspective, little has been done.

As is reflected in the preceding quotation, to a casual observer and commentator, an interactional perspective may seem obvious, even trivial. However, when taken seriously, it has far-reaching and important implications for planning, implementation, and interpretation of psychological theorizing and research. This volume is an attempt to emphasize these implications for developmental research and to take these implications into account during actual empirical investigations of developmental problems.

Interaction is a central principle in functioning of open systems at all levels, from the macrocosmos to the microcosmos (cf. von Bertalanffy, 1968; Bronfenbrenner, 1979a; Miller, 1978). Components of open systems do not function in isolation, and they usually do not function interdependently in a linear way. The interaction is much more complex, particularly for the biological and psychological systems which interact together in an individual. Within the biological system, interdependence and non-linear interrelationships reflect a fundamental principle. The same is true for the relationship among the various aspects of the psychological system that are defined in terms of hypothetical constructs. Interaction is also the fundamental principle underlying the relationship between an individual and the environment.

Three Basic Propositions

The interactional perspective for individual functioning rests upon three basic propositions that should be considered simultaneously:

1. The individual develops and functions as a total, integrated organism. Development does not take place in single aspects per se, in isolation from the totality.
2. The individual develops and functions in a dynamic, continuous and reciprocal process of interaction with his/her environment.
3. The characteristic way in which the individual develops, in interaction with the environment, depends on and influences the continuous reciprocal process of interaction among subsystems of psychological and biological factors.

The meaning and implications of these three propositions are discussed in the following sections with special reference to aspects that have relevance for the planning and interpretation of the longitudinal project.

A Holistic "Person" Approach

In general, psychological research in fields such as intelligence, cognition, learning, information processing, decision making, and motivation is *variable oriented*; that is, the variable is the main object of interest. Empirical research in this tradition is often closely bound to a specific theoretical proposition and directed to the study of hypotheses derived from the theory with respect to the variables under consideration. Theories are often tested concerning simple, usually noninteractive relationships among a few variables in a mechanistic, reductionistic measurement model. The planning of the collection and analysis of data is often so bound to the testing of a specific theory that the data obtained are difficult to use for other purposes or to interpret in other frames of reference. This orientation of research toward variables has been in general fostered by the development of effective methods for variable analysis and hypothesis testing such as analysis of variance, regression analytical methods and factor analysis.

Even if the tendency to make the variable the main conceptual unit of analysis is less conspicuous in developmental research than in general psychology, the same tendency is also present. Consequently, the nature of the relationship among variables, in terms of S-R relations or R-R relations, becomes the main object of interest also in this field. The individual is then important in providing measures for the variables.[2]

The variable approach to developmental research is closely tied to a reductionistic model of man (see Overton & Reese, 1973; Pepper, 1942). Scarr (1981) distinguished three forms of reductionism in behavioral genetics and developmental research, of which two, reduction in the level of explanation and reduction in the methods of investigation, can be applied to much variable-oriented research. Reduction in the level of explanation is, according to Scarr, "that one appeals to phenomena that are parts or constituents of the phenomena one wants to explain without specifying how other parts are organized into the whole" (p. 163). Reduction in the methods of investigation is described as "the analysis of phenomenon into bits and pieces that seldom are reassembled in a satisfactory explanation of the original phenomena" (p. 163). The approach that these formulations reflect has ancient roots and has had a dominant influence on psychological research, particularly in the experimental S-R tradition. The tendency to break down psychological structures and processes into the smallest possible pieces was discussed and criticized by James (1890). Dewey (1896), following James' criticism of the atomistic approach to mental thought processes, warned that the S-R approach could imply another form of atomism. It is interesting to observe, 100 years later, the actual direction taken by most forms of psychological research.

An interactional view emphasizes an approach to the individual as an organized whole, functioning as a totality and characterized by the partially specific patterning of relevant aspects of behavior. This view can be designated a "person" approach to the study of psychological phenomena.

Thus, the point of departure for the person approach is the proposition that an individual functions as a totality and that each aspect of the structures and processes that are operating (perceptions, plans, values, goals, motives, biological factors, conduct, etc.) takes on meaning from the role it plays in the total functioning of the individual: "There is a logic and coherence to the person that can only be seen in looking at total functioning" (Sroufe, 1979a, p. 835). The substantive background to this statement was formulated by Kuo (1967) in his

---

[2]One of the most misused terms in psychology is "variable." It was originally defined in mathematics as "a quantity that may assume any values or set of values". The meaning of the word was extended in experimental research to mean "a factor which may vary, e.g., in an experiment." In common psychological language, the use of the term is closely connected with the tendency to reification of psychological phenomena that are basically inferential in nature, but do not exist as anything other than a specific aspect of the functioning of the total organism, viewed from a certain perspective. The organism as a totality can function in an intelligent, dependent or helpless way, but intelligence, dependence, and helplessness do not exist per se. The use of the term variable is here restricted to cases where a certain aspect has been defined and operationalized in a certain way.

discussion of what he called behavior gradients: "The most essential feature of the behavioral gradients concept is that, in any given response of the animal to its environment, internal or external, and in any given stage of development, the whole organism is involved" (p. 92). The same can be formulated for human individual functioning, implying that the individual as a whole must be considered in the planning and interpretation of empirical research on specific aspects of individual functioning. The whole picture has an information value that is beyond what is contained in its separate parts (the "doctrine of epigenesis"): "Behavior, whether social or unsocial, is appropriately viewed in terms of an organized system, and its explanation requires a 'holistic analysis'" (Cairns 1979a, p. 325).

The holistic approach to psychological research is not new. It has been discussed at length by prominent researchers interested in psychological phenomena in a current perspective (cf. Allport, 1937; Lewin, 1935; Russell, 1970; Sroufe, 1979a) as well as by researchers mainly interested in a developmental perspective (cf. Block, 1971; Cairns, 1983; Sameroff, 1975, 1983; Wapner & Kaplan, 1983; Wolff, 1981). (See also a recent review of Soviet psychology, Asmolov, 1984.) The issue about person vs. variable approaches is reflected in the debate over idiographic vs. nomothetic, typological vs. dimensional, and clinical vs. statistical approaches to empirical psychological research. Most of the time, the two approaches have been regarded as contradictory. In later chapters of this book it is argued that they are compatible, and that what superficially seems to be contradictory in methodology and empirical results is often the result of inadequacies in theoretical and conceptual distinctions and in methodological sophistication.

PERSON-ENVIRONMENT INTERACTION

The second of the three propositions underlying the interactional approach, which emphasizes the dynamic, continuous, and reciprocal process of interaction between the individual and the environment, is the fundamental feature of what might be called *classic interactionism* (see Ekehammar, 1974; Endler, 1983, 1984; Endler & Magnusson, 1976a; Magnusson, 1976; Magnusson & Endler, 1977b).[3]

---

[3]The view reflected in this proposition has been advocated and discussed by many researchers during recent years under various headings. For example, Pervin (1968) adapted the term "transactionistic," and Bandura (1978) the term "reciprocal determinism" for individual functioning in current terms. Baltes, Reese, and Lipsitt (1980) used the term "dialectic-contextualistic," Bronfenbrenner and Crouter (1983) the term Process-Person-Context model, and Lerner and Kauffman (1985) the term "developmental contextualism" for individual functioning in a developmental perspective.

A "Current" Perspective

The debate about an interactional view during the last decades has been mostly concerned with individual functioning in a "current" perspective. The main points of such a view were summarized by Endler and Magnusson (1976a, p. 4) as follows:

> 1. Actual behavior is a function of a continuous process of multidirectional interaction of feedback between the individual and the situations he or she encounters.
>
> 2. The individual is an intentional, active agent in this interaction process.
>
> 3. On the person side of the interaction, cognitive and motivational factors are essential determinants of behavior.
>
> 4. On the situation side, the psychological meaning of situations for the individual is the important determining factor.

Behavior cannot be understood in isolation from the environmental conditions in which it occurs. This proposition has long been advocated and elaborated by researchers with very different perspectives: behaviorists, phenomenologists, personologists, social psychologists, trait psychologists, and those advocating a psychodynamic view (Angyal, 1941; Brunswik, 1952; Cattell, 1963, 1965; Jessor, 1956; Kantor, 1924, 1926; Kelly, 1955; Koffka, 1935; Lewin, 1935; Murray, 1938; Rotter, 1954, 1955; Sells, 1963, 1966; Staats, 1963; Tolman, 1949; Vernon, 1964). One of the most explicit of these formulations was that of Brunswik, who suggested that psychology be defined as the *science of organism-environment relationships*. The importance of considering the role of environmental factors in models of behavior has also been underlined by researchers in neighboring disciplines such as sociology and anthropology (Arsenian & Arsenian, 1948; Berger & Luckman, 1966; Chein, 1954; Goffman, 1964; Mead, 1934; Thomas, 1927, 1928).

The view that environmental factors, particularly those in the immediate situation, influence behavior and should be considered in models for individual functioning has to be qualified by a distinction between *general* situational effects and *differential* situational effects (Magnusson, 1984b).

General situational effects are the same for all individuals. It takes longer for all individuals to climb a high mountain than a low one, and most individuals would react more strongly if they met a tiger, rather than a house cat. In this case, when cross-situational instability in behavior is assumed to be due to general environmental factors only, cross-situational profiles for a sample of individuals with respect to a certain type of behavior will show stable rank orders of individuals. A person-situation matrix of data for a specific aspect of individual functioning, such as adrenaline excretion in the urine, aggressive behavior, or achievement of tasks at differing levels of difficulty, will then contain only one source of individual variance: the main variance due to persons.

The notion of differential situational effects is that, in addition to the general situational effects, there are also effects that are specific for individuals or groups of individuals (Lewin, 1931). The individual differences in this respect are assumed to be dependent on the individuals' partially specific interpretations of stimuli and events in the environment. For example, some people experience situations characterized by strong demands for achievement as threatening and react negatively, psychologically and/or physiologically, while others experience such demands as challenging in a positive sense and react accordingly. On the other hand, the latter group may become threatened and react with anxiety to situations involving expectation of separation or physical pain (Ekehammar, Schalling, & Magnusson, 1975). Assuming both general and differential situational effects an individual's profile for a certain type of behavior in different situations will have two characteristics: a mean level and a partially specific cross-situational form. A person-situation matrix of data will contain two sources of individual variance: the main variance due to persons and the interaction variance, which is composed of the individually specific part of the cross-situational profiles. Thus, individuals differ, in terms of data, with respect to their cross-situational *patterns* of behavior. That such patterns are stable over time has been empirically demonstrated by Magnusson and Stattin (1981c).

Researchers, referred to as interactionists in their formulations, have not always been clear about the distinction between general and differential situational effects and about the consequences for methodology and research strategy. However, if the strong theoretical formulations about the role of environmental conditions for individual functioning are to have meaning beyond the obvious and trivial, particularly with respect to implications for psychological research, they must imply the operation of differential situational effects, as reflected in Lewin's well-known formula: B=f(P, E). However, with respect to progress in psychological theory and empirical research, it is important to note the limited impact that interactional formulations have had on empirical research, despite the strong emphasis on environmental factors by most researchers and the very strong theoretical, methodological and strategical consequences of the existence of differential situational effects. The following comprise two striking examples.

The first example is the dominant *measurement model* (as distinguished from *theories* about psychological phenomena) frequently applied in personality and much developmental research. Much research applying this measurement model uses nonsituation-specific rather than situation-specific data, an approach that implies the non-existence of differential situational or environmental effects. This measurement model and the methods closely connected with it have had a strong impact on personality research during the last 30 to 40 years. (For a further discussion of the consequences of the application of this measurement model, see Magnusson, 1976, 1984b.)

The second example is from research on stress and anxiety. In this area the integral role of environmental situational conditions has been particularly emphasized and incorporated into theoretical models (Endler, 1975; Lazarus & Launier, 1978; Rutter, 1983; Spielberger, 1977). Thus, one would expect this to be a field in which differential situational factors are particularly taken into account in empirical research. In order to examine the impact of these theoretical formulations on empirical research, articles on stress and anxiety in two leading journals, *Psychosomatic Medicine* and *Psychophysiology*, were examined for the period 1970–80 (Magnusson, 1984a). A total of 57 studies came under the heading of stress and anxiety during the sample period. In no case was there a systematic sampling of situational conditions to investigate or control possible differential effects on the subjects' stress and anxiety reactions. In 31 studies the stress-inducing factor was threat of physical pain, generally expectation of a mild electric shock. In only two studies did the researchers discuss the possible implications of choosing a specific type of situation for the experiment. This was generally true as well for the studies concerned with gender differences in stress and anxiety reactions. Given the empirical evidence for strong differential situational effects reflected in sex differences, age differences within sexes, and individual differences within sexes and ages with respect to reactions to stress and anxiety-provoking environmental and situational conditions, the negative consequences of neglecting possible differential situational effects in planning and interpreting results in this area become obvious (Magnusson & Olah, 1981).

This review of research in the field of stress and anxiety illustrates the limited impact that strong theoretical formulations concerning the role of environmental conditions have had on empirical research. This limitation holds for many other areas of personality and psychological research as well. In developmental research there has been much discussion about the role of environmental conditions and the interaction of person and environmental factors. However, there has been more talk than scientific endeavor (Cairns & Valsiner, 1984).

A Developmental Perspective

The current functioning of individuals reflects the influence of their past course of development. The individual's readiness to interpret and respond to a certain situation has been formed by continuous interaction with various situations in the past. The environment provides information to process and offers necessary feedback for the building of valid conceptions of the outer world as a basis for interaction. In the continuous interaction with the environment in its physical, cultural, and social manifestations, individuals develop a total integrated system of mental structures and contents that shape and constrain their methods of functioning. On the basis of and within the limits of inherited dispositions, affective tones become attached to specific contents and actions, and strategies are developed for coping with various kinds of environments and situations.

The fact that the course of development is dependent on the person's past experiences of physical, social, and cultural characteristics of the environment is common knowledge. This second proposition of the interactional perspective has been emphasized by researchers interested in the issue of individual development. As Cairns (1980) reports, as long ago as the 1890s, Baldwin explicitly discussed ontogenetic and evolutionary development in interactional terms (Balddwin, 1895, 1897). And, as suggested by Cairns and Cairns (1985), there is a direct line from Baldwin to Piaget's (1928) work on language and thought, Kohlberg's (1969) studies on ethical development, and others who have influenced various areas of developmental research. During the last decade, interactional formulations of individual development that stress the role of environmental conditions for the developmental process, have been presented by Flavell (1982) in cognitive development, Ulvund (1980) and Nygård (1984) in cognition and motivation, Scarr (1981) in behavior genetics, Cairns (1976, 1977, 1979a) and Dunn (1981) in social development, Levine (1982) in psychobiology, and Lerner (1978) and Thomas and Chess (1977, 1980) in temperament, among others. An important contribution to the unification of developmental research in interactional terms was Sameroff's (1975, 1982) application of a systems theory (see also Urban, 1978).

These examples demonstrate that the interactional approach to development as a process characterized by an interplay between persons and environmental factors has been more widely accepted as a framework for research on individual development than, for example, for research on personality. Nevertheless, the important theoretical, methodological, and strategical implications for developmental research of an interactional perspective have not yet been realized to the extent they deserve (Lerner, Skinner, & Sorell, 1980).

The Environment in Person-Environment Interaction

Most directly, individuals meet their environment in specific situations. In actual physical terms a situation can be defined as that part of the environment that is accessible for sensory perception at a certain occasion (Magnusson, 1981b). Within the frame of each situation, stimuli and events that influence an individual's behavior and that are influenced by the individual change constantly. The actual situation and the specific situational stimuli, cues, and events are interpreted by the individual, and the meaning assigned to the total situation provides the main basis for the interaction with the environment. As has been stressed by many researchers, the environmental influence on individual current functioning and development is not limited to the immediate situation. Instead, the immediate situation is embedded in a larger environment with physical, social and cultural properties operating both directly and indirectly at all levels of specificity-generality in the person-environment interaction (Barker, 1965; Bronfenbrenner, 1977, 1979a, b; Jessor, 1981; Magnusson, 1981b; Pervin, 1978; Takala, 1984). At all those levels, aspects of the environment can influence the individual's current behavior and developmental life course.

The environment as a source of stimulation vs. a source of information. When we discuss environmental influences on the development of the individual, an important distinction should be made between the environment as a *source of stimulation* and the environment as a *source of information*.

This discussion relates to the old distinction between the external world "as it is" and as it is perceived and subjectively interpreted to be. For the environment as it is, Koffka (1935) introduced the term "geographical environment" and Murray (1938) talked about "alpha situations." The perceived environment was referred to by Koffka (1935) as the "behavioral environment," by Lewin (1935) as "life space," by Murray (1938) as "beta situations," by Tolman (1951) as "the immediate behavior space," and by Rotter (1955) as "the psychological situation" (cf. also Wohlwill's distinction between the environment as a source of active stimulation and as a "context of behavior," 1973). The actual, physical environment acts upon the individual in important respects that can be reacted to without an intermediate process of interpretation. For example, this effect occurs for some biological reactions to environmental stimulation as well as for fear reactions to physical stimuli, as suggested by Zajonc (1984), or when the organism reacts to certain viruses (cf. Öhman, in press). The view of the environment as a source of stimulation that elicits and releases individual responses has dominated respected fields of psychology for decades, with far-reaching consequences both for theory about psychological phenomena and for planning and interpretation of empirical research.

Illustrations of the approach to the environment primarily as a source of stimulation can be found in traditional experimental psychology and classical learning theory. The main stream of research in the flourishing field of environmental psychology has been based on this view. In experimental psychology, individual responses are studied as a function of external stimulation under the assumption that the stimuli are interpreted in exactly the same way by all subjects. In classical learning theory, emphasis is placed on the establishment of S-R contingencies in which responses are "stamped in" as reactions to stimuli and events.

The importance of considering the environment as a carrier of information has become obvious and is explicitly discussed in psychological formulations such as the following by Aronfreed (1968, p. 68): "One of the points of agreement among contemporary versions of general behavior theory is on the requirement that direct stimulus control must be subordinated to representational cognition" (see also Kagan, 1967). According to this view, the most important role of the environment is to offer the information that makes it possible for the individual to understand the world around him and himself in relation to the world:

> "The essential point is that the information thus received, coordinated, and integrated into some kind of representation provides the effective stimulus for the instigation and guidance of goaldirected behavior. If such a cognitive representation does exist, one of its adaptive functions is that the same cognitive representation may elicit a variety of motivations depending upon the state of the organism, and may guide a variety of behavior sequences directed toward different targets. In other words, the individual cognizes his environment in much the same way whether he is hungry and utilizes this information about the environment to find food; whether he is bored and utilizes this information to find entertainment; or whether he is motivated to go to work and utilizes the cognitive representation of the environment to guide him to his office." (Baldwin, 1969, p. 328)

In order to understand the way an individual interacts with the environment at various levels of complexity, one has to accept the assumption that the environment essentially serves as a source of information. This view is reflected in modern social learning theory that assumes that an individual's way of dealing with the external world develops in a learning process in which two types of perceived contingencies are formed. These are: (a) *situation-outcome contingencies* (implying that certain situational conditions will lead to certain outcomes), and (b) *behavior-outcome contingencies* (implying that certain actions by the individual will have certain predictable consequences) (cf. Bolles, 1972). The formation of situation-outcome and behavior-outcome contingencies constitutes one source for the stability and continuity of individuals' functioning in relation to the environment. Such valid contingencies enable the individual to make predictions about the external world, to exert *predictive control*. In addition, the formation of such contingencies allows the individual to foresee the outcomes of different lines of action and to use that knowledge as a basis for effective purposive action, to exert *action control*. The individual's ability to predict and take active control of the environment forms the fundamental basis for goal-directed activity and for the experience of meaningfulness (Brandtstädter, 1984; Mineka & Kihlstrom, 1978; Weisz, 1983).

The essence of this view was well formulated by Sells (1966):

> "Adaptive function of an organism also implies the existence of feedback mechanisms. The posture of the organism at any moment is in effect the expression of an intrinsic (and not necessarily consciously experienced) *hypothesis* concerning the nature of the environment. Every response is similarly an *interrogation* of the environment and the resulting feedback provides information (also not necessarily conscious) that enables adaptive response. The existence of biologic and neurophysiological feedback systems is a necessary assumption about adaptive organisms." (p. 133)

The importance of this process in early development was emphasized by Lewis and Brooks-Gunn (1979) when they suggested that "the first definition or feature of self is the simultaneity and identity of action and outcome" (p. 224).

In this perspective, two basic requirements of an optimal environment for individual development are that it should be both *consistently patterned* and *influenceable* (Magnusson & Allen, 1983b). Both conditions are necessary in order for the individual to learn how to handle the environment and to deal with it in current situations, that is, to make the environment meaningful as a frame for purposive actions. It is not enough that the environment is patterned and consistent; it must also be possible for the individual to control it through his/her own actions. There is an important difference between physical and social environments in these two respects (cf. Wohlwill, 1981). The physical environments usually function in a rather coherent and stable manner and are generally easily interpreted by our senses. The arrangement of the physical environment, as well as the variety of stimulation and information it offers, has implications for the development of sensory perception (cf. Hubel & Wiesel, 1970) and for the cognitive development of children (cf. Bradley & Caldwell, 1976; Hebb, 1958; Hunt, 1961, 1979, 1981; Yarrow, Rubenstein, & Pedersen, 1975).

Social environments have no inherent patterns. Instead, the developmental process is influenced by the patterning of social environments that is created by other people, particularly those who take care of the children. The extent to which the social environment is meaningfully structured as well as how it is structured, that is, the system of values, norms, rules, and roles, underlying the behavior of significant persons in the environment, is essential in the development of the child's conceptions of the external world and his/her role in it. It is the patterning and consistency of other people's behaviors, such as their demands and their rewards and punishments, that help the child develop a sense of order and lawfulness which can be used to assign meaning to the environment and to make valid predictions about situation outcome contingencies and behavior outcome contingencies in the external world.

The view described in the preceding section can resolve the dispute between Emmanuel Kant's idealistic idea that our consciousness is not shaped by reality, but reality is shaped by our consciousness and Karl Marx' materialistic standpoint that reality is not shaped by our consciousness, but our consciousness is shaped by reality. According to the view advocated here, both are right; consciousness is formed in a continuous process of interaction with reality and is thus dependent on it; at the same time consciousness defines the reality which forms the basis for individuals' purposive dealing with the environment.

Other People as an Important Aspect of the Environment

In the preceding section other people were presented as the most significant aspect of an individual's environment, both in a current and in a developmental perspective (Shantz, 1983). The role of other people in the dynamic person-environment interaction was emphasized by Patterson and Moore (1978), who defined interactionism in terms of person-to-person interactions. Person-to-person interactions in a developmental interactional perspective were discussed by

Peterson (1968), who cogently discussed the role of environmental factors for individual functioning (see also Peterson, 1979). McClintock (1983) discussed theoretical and empirical gains that can be obtained from interactional analyses of social relationships. Parents, siblings, teachers, peers, and others play a crucial role in the development and establishment of the perceptual-cognitive representations and conceptions of the external world, thereby providing a base for interpretation of the information offered in current situations. Thus these persons in the proximal upbringing environment and the way they structure the social environment for a child are important in the socialization process through which an individual learns and integrates into his own personality the values, norms, roles, and rules of his culture. The degree to which these other people are "successful" in the socialization process influences the individual's adaptation to society. The central role of family management practices and of parental skill and discipline in that process was recently emphasized by Patterson (1986) when he discussed three models for the development of antisocial behavior among boys. The interesting and important issue of gender differences in parents' practices and in the effect of family factors has been investigated by Hetherington and her colleagues (see Heatherington, Cox, & Cox, 1979), among others.

Self-perceptions

A central aspect of an individual's mental life (given the assumption that the environment serves primarily as a source of information) involves self-perception and self-evaluation based on the total conception of the external world. (James devoted an entire chapter to this issue in 1890.) In a current perspective, these perceptions and evaluations play an important role in the selection and interpretation of information from the external world, in the individual's sense of control over his future, in his conduct in current situations, in the way he relates to other people (e.g., trustful or suspicious), and so forth. The development of the individual's self-perception, self-evaluation, and self-respect forms a main element in the learning process during which the individual learns to exert predictive and action control over the environment (Bandura, 1978; Brandtstädter, 1984; Harter, 1983; Weisz, 1983).

The role of the social environment in this process, with its values, norms, rules, rewards and punishments, has been emphasized by many researchers over the last century (see Baldwin, 1897, who proposed that the self in its various aspects is essentially a product of social processes; Cooley, 1902; and Mead, 1934). Lewis and Brooks-Gunn (1979) stated that the self ". . . . is developed from the consistency, regularity, and contingency of the infant's action and outcome in the world" (p. 9). Central to the discussion regarding the development of self perceptions and their role in current functioning are intelligence and mental capacity in its various aspects (Gardner, 1983), subjective competence (Bowerman, 1978; Heckhausen, 1983), inner and outer control (Rotter, 1955), predictive and behavior control (Mineka & Kihlstrom, 1978), motivation (McClelland, 1955) and learned helplessness (Abramson, Seligman, & Teasdale, 1978; Seligman, 1975). An important aspect of this process of self-perception and evaluation is the effect on the individual's sense of control over the environment; for example, one's ability to influence the direction of one's future educational and vocational career.

Reciprocity in Person-Environment Interaction

A characteristic feature of the environmental approach to individual functioning, as discussed earlier, is the unidirectional causality, that is, the assumptions that environmental properties determine individuals' ways of functioning. This tradition is closely linked to a behaviorist view, as expressed by Skinner: "A person does not act upon the world, the world acts upon him" (Skinner, 1971, p. 211). The measurement model that is usually applied to data in this tradition reflects a mechanistic relation with linear, additive effects on behavior of environmental factors (cf. Magnusson, 1976).

Interesting examples of the application of this view in empirical research can be found in traditional research on the role of social relations for the developmental process and individual well being. For a long time the role of parents in the development of a child was investigated mostly in terms of the parents' influence on the child. A common measure of children's social relations is still the *number* of peers who like, reject, or neglect them. The measurements are then used for the study of the relationship between social factors in models which regard the number of peers as the independent variable and the child's social adjustment as the dependent variable. The same approach holds for the vast amount of recent research on social networks and their role in individual satisfaction and ability to deal with adversity. In these cases the reciprocal character of relationships is often neglected. The extent to which peers and friends contribute positively or negatively to individual satisfaction and behavior depends on a number of factors. For example, Riley and Eckenrode (1986) found that social ties can be stressful as well as supportive. Women with greater material and psychological resources derived more beneficial support and suffered less stress from their social ties than women with fewer resources in these areas.

A central aspect of the interactional perspective is the reciprocal character of the person-environment interaction process (Bandura, 1978; Bronfenbrenner & Crouter, 1983). Reciprocity implies that the individual both influences and is influenced by the environment at each stage of development. In this reciprocal interaction process, the individual is not a passive receiver of external stimulation but rather an active and intentioned player who interprets information about environmental conditions and events and acts in the frame of reference of his own systems of thoughts, values, goals, and emotions. The individual influences the environment according to his own goals and motives and plans, seeking some situations and avoiding others (Bandura, 1980; Lerner & Busch-Rossnagel, 1981). As underscored earlier, an important role in the person-environment interaction process is then played by the meaning that individuals assign to their environments.

But the individual does not act upon a passive environment, as is implicit in some influential developmental theories (cf. Riegel, 1978). The environment acts on the individual in a way that is often initiated and maintained by the actions of the individual. The best illustration of the reciprocity in person-environment interactions can be drawn from person-person interaction, particularly parent-child interaction (cf. Bell, 1968; Davis & Hathaway, 1982; Hartup, 1978; Maccoby & Jacklin, 1983; Kessen, 1979; Parke, 1978; Sears, 1951). To a certain extent, the behavior of one individual influences the behavior of others, whose behavior in turn will affect the reactions and actions taken by the first, and so on. A child influences the behavior of the parents and other family members who form an important aspect of its own developmental environment; the child is both the creation of and the creator of his or her environment. The relevance of this view is well illustrated in the empirical study presented in Chapter 6.

"Life Events" and "Chance Events" in Person-Environment Interaction

One issue that has drawn much interest both in theory and empirical research comes under the heading of "life events." Most of this research has been devoted to the role of important life events in the physical and mental well-being of the individual (Dohrenwend & Dohrenwend, 1974). (Theory and empirical research on the role of life events in the life span development perspective have been reviewed by Brim and Ryff, 1980, and Hultsch and Plemons, 1979, among others). Another type of event that has drawn less attention, but that may have far-reaching effects on individual development, is what Bandura (1982) called "chance events."

One of the main problems with open, multi-determined systems is balancing the operating factors so as to maintain a steady state. Operating factors may balance each other in the ongoing interaction process to the extent that under normal conditions the process is insensitive to external factors. However, on some occasions the system may be unstable and otherwise vulnerable, so that even a small external influence might change the direction and strength of the process. This general view can be applied to the course of development of individuals as well as to organizations at various levels of complexity.

The developing individual can be considered as an open, multi-determined system, that is generally stable and insensitive to the influence of external factors, even if they occur frequently and with great strength. It is only at certain points in time and under certain conditions that the system is affected in a major sense, and a more conspicuous change will take place. This implies, not only maturational timing effects or "readiness" factors, but also chance events that may change the course of development at crucial junctures. Major events or seemingly small events may have profound effects on a person's life course if they occur at a certain stage of development. On the other hand, under other conditions, or during other stages, the same event might have no effect at all. Bandura (1982) gives striking examples of how events that seemingly occur by chance can have decisive effects on an individual's life path. Such incidents often happen in an individual's "choice" of educational and vocational career. It makes many theories that assume that decisions in these respects are made on logical grounds less realistic. The occurrence of chance events, sometimes of great importance for an individual's life path, also has implications for the discussion about prediction as a goal for psychological research (cf. Lipsitt, 1983). This issue is discussed in the next chapter.

An important aspect of the role of "chance events" in the development process should be observed. The chance occurrence of certain situations at crucial points in individuals' courses of development does not imply that the effect on individual development is random. The randomness of the occurrence of such situations is on the environmental side of the person—environment interaction process, and "chance" refers to what happens in the environment. However, if and to what extent situations that appear more or less by chance have any effect on individuals, depends on them and on whether or not they observe the opportunities or risks offered by the situations, are prepared to take advantage of possible opportunities, and so on.

The Person: An Active Purposeful Agent

A central aspect in the view of individual functioning discussed in the preceding sections is that the individual is not only a passive receiver of stimulation from the environment to which he reacts; he is an active purposeful agent in the person-environment interaction process. One important implication of this facet of interactionism is that when the process underlying individual functioning and development is described in terms of person-situation or person-environment interaction, the central issue for psychological theorizing and research is not how the person and the environment, as two separate parts of equal importance, interact. Rather, and this is essential, it is how individuals by their perceptions, thoughts, and feelings function in relation to the environment. Thus in individuals' dealing with the external world, a fundamental role is played by their integrated mediating system of which the main aspects are their cognitions and conceptions of the external world (including their self-perceptions), their way of processing information, their emotions, their goals and values, and their physiological processes. The structure and functioning of an individual's mediating system is formed and changes slowly in a process of maturation and experience that takes place in the continuous, bidirectional interaction between the individual and the environment.

It is this mediating system that determines which situations individuals seek and which they avoid (as far as they have options), which situational conditions they attend to, how they interpret single stimuli and patterns of stimuli and events, and how they transform the information that the environment offers into inner and outer actions. Understanding the lawfulness of how subsystems of perceptions, cognitions, goals, values, emotions, and physiological processes function in interaction with each other and as an integrated total system in current situations and how these subsystems develop during the process of maturation and experience in development, therefore becomes a central task for psychological theorizing and research.

In the mediating processes that characterize a person's purposeful dealing with his environment, a central, and in some connections a decisive, role is played by basic values and norms that are relevant for the particular issue under consideration. The value structure underlies and determines the long-term and the short-term goals that steer an individual's thoughts and actions in many situations (Pervin, 1983). The sometimes very strong impact of values on individuals' way of functioning in some situations can be easily seen in the history of politics and religion. It is surprising how relatively little psychological research has been devoted to the role of basic personal values in individual long-term functioning as well as in specific situations. One of the most important aspects of education in the family and at school is the transfer of social values from one generation to the other. For developmental research the role of values in the socialization process and the role of various agents in the transfer of values and norms to the youngsters should be one of the most central issues.

The view summarized above has important implications, one of which was expressed by the Nobel prize laureate Sperry (1982) in the following way:

> "The key development is a switch from prior non-causal, parallelist views to a new causal, or "interactionist" interpretation that ascribes to inner experience an integral causal control role in brain function and behavior. In effect, and without resorting to dualism, the mental forces of the conscious mind are restored to the brain of objective science from which they had long been excluded on materialist-behaviorist principles.
>
> The spreading acceptance of the revised causal view and the reasoning involved carry important implications for science and for scientific views of man and nature. Cognitive introspective psychology and related cognitive science can no longer be ignored experimentally, or written off as "a science of epiphenomena" or as something that must in principle reduce eventually to neurophysiology. The events of inner experience, as emergent properties of brain processes, become themselves explanatory causal constructs in their own right, interacting at their own level with their own laws and dynamics." (p. 1226)

INTERACTION AMONG SUBSYSTEMS IN THE INDIVIDUAL

The third basic proposition of an interactional perspective is that the characteristic functioning of an individual in a dynamic person-environment interaction depends on and influences the continuous reciprocal process of interaction among psychological and biological subsystems. As emphasized earlier, the first two propositions of an interactional perspective have become standard formulations. This is not the case for the third proposition; it belongs to what might be called "modern interactionism." It has its roots in recent developments in neuropsychology, endocrinology and pharmacology, which have rapidly illuminated the continuous interplay between biological brain processes, other physiological processes, mental processes (thoughts and emotions) and behavior. These findings help bridge the gap between biological and psychological sciences and contribute to a better understanding of psychological processes.

The research in these areas demonstrates the importance of biological factors for current functioning and for the developmental process and emphasizes the need to incorporate biological factors into psychological models in order to understand why individuals think, feel, act, and react as they do. The role of biological factors in psychological phenomena was recognized early in the history of psychology. This was not surprising, given the fact that some of the fathers of psychology had their roots in physiology. In 1883, Wundt made a plea for psychology as an independent scientific discipline that emphasized the biological base for psychological phenomena: "The new discipline rests upon anatomical and physiological foundations, which in certain respects, are themselves very far from solid." (Wundt, 1948, p. 248). Since the beginning of the history of differential psychology, the role of genetically determined factors in the course of individual development has been a main issue (cf. Galton, 1865; 1869). In general terms, the importance of the interplay of psychological and biological factors has been emphasized in psychosomatic medicine (Carlsson & Jern, 1982; Jenkins, 1985). However, in developmental research, the role of biological factors in the total intrapersonal processes has been generally neglected. Though the role of biological factors in the development process was emphasized by some researchers, the interactional character of the processes in which mental and physiological factors are involved has only recently been studied. It is worth noting that few of the models which emphasize development as a dynamic, person-environment interaction process include or discuss the possible role of biological factors in that process. Exceptions can be seen in the work of Bronfenbrenner and Crouter (1983), Cairns (1979a), and Lerner (1984) and of psychobiologists, such as Gottlieb (1983), Kalverboer and Hopkins (1983) and Levine (1982). One of the main drawbacks of much traditional developmental research is that it has to a large extent neglected to consider biological factors in the planning and interpretation of empirical research on developmental processes.

Some psychologists who are against this position argue that they are psychologists, not physiologists. Of course, the central issue is not to understand physiological processes; it is to understand individuals' thoughts, feelings, actions, and reactions. When the interaction between mental and biological factors in terms of reciprocal and dynamic interaction processes is discussed, the contention is similar to that made earlier concerning the interaction between the individual and the environment; in psychology the matter is not two subsystems of equal interest. The main interest is why individuals think, feel, act, and react as they do. But to say that this focus implies that biological processes should not be considered in the search for an explanation of individual functioning is as if a meteorologist were to claim disinterest in how the landscape is shaped, as he was only dealing with the climate.

Biological and Cognitive-Affective Factors

Mental life, thoughts, and emotions are influenced by bodily conditions. The dependence of thoughts and emotions on physiological processes has been elucidated in much empirical research (Öhman, in press). It has demonstrated, for example, that moods are dependent on the excretion of serotonin and the metabolite 5HIAA in the plasma. A low level of serotonin results in depression and has been found to be related to suicidal behavior (see Åsberg, Mårtensson, & Wägner, 1986; Åsberg, Schalling, Rydin, & Träskman-Bendz, 1983). (See Lerner, 1984, among others for more comprehensive reviews).

However, the relationship between biological and mental factors is not generally unidirectional; it is also reciprocal. Physiological processes can be evoked by cognitive-affective events and are maintained in a continuous process of interaction between mental and biological factors. Of particular interest for an interactional model of individual functioning is the essential role played by the individual's interpretation of the environment. The appraisal of external information guides thoughts and actions and evokes physiological systems that in turn influence psychological events, thoughts, and emotions. Certain bodily systems that characterize a person are particularly sensitive to the challenge of specific situational conditions. A good illustration is the reaction of individuals to situations appraised as threatening. Such an appraisal leads, among other things, to the activation of the autonomic nervous system and the excretion of adrenaline from the medulla. The activation of this physiological system influences other bodily processes, and the total effect of these changes is to affect the individual's emotions and thoughts. A series of experiments on monkeys by McGuire and his coworkers have demonstrated that social factors, such as the status of the leader in the group and his interpretation of the behavior of other group members affect his level of serotonin and 5-HIAA (McGuire, Raleigh, & Johnson, 1983a,b; Raleigh, McGuire, Brammer, & Yuwiler, 1984). These examples illustrate the dynamic, reciprocal character of the continuous interaction among biological and psychological subsystems of an individual and the role of environmental factors in that process. Interest in this area has led to the formulation of *sociopharmacology* as a special field of research (McGuire, Raleigh, & Brammer, 1982).

Of interest for further discussion is the gender difference found in adrenaline excretion in situations involving stress and anxiety. In situations where stress reactions are induced by demand for achievement, males excrete significantly more adrenaline than females, while the tendency is the opposite for situations involving a possible threat to the parent's child. A reasonable interpretation of this empirical finding is that gender is responsible for different reactions in these two situations (Magnusson, 1984a).

THE INTERPLAY OF GENETIC AND ENVIRONMENTAL FACTORS

With an unusual unanimity, empirical studies demonstrate a positive, significant correlation between parents and children with respect to intelligence, which indicates a genetic background to individual differences in intelligence. In the area of temperament, Buss and Plomin (1984) emphasized that interindividual differences are determined to some extent by genetic factors. Studies of rodents, dogs, and monkeys have shown that individual differences in aggressive behavior can be strengthened and established by selective breeding (Cairns & Nakelski, 1971; Lagerspetz & Lagerspetz, 1971). On the basis of analyses of questionnaire data from monozygotic and dizygotic pairs of twins, Rushton, Fulker, Neale, Nias, & Eysenck (1986) drew the conclusion that interindividual differences in altruism and aggression are substantially inherited. In a traditional biological approach to individual functioning, as briefly described previously, these results would be interpreted as examples of a unidirectional, mechanistic influence of genetic, inherited factors on individual functioning. In an eloquent refutation of such a reductionistic view, Scarr (1981) states that:

> "Genes are but constituents in larger systems. Gene action occurs in the context of cells and reverberates throughout the higher levels of organization, including behavior. But there are always concurrent nongenetic events. Developmentally, gene action both initiates growth and is regulated by the growth of other constituents in the system. . . . . Probabilistic models do not expect any one-to-one correspondence between genotype and phenotype. Rather, they emphasize the interplay of untold numbers of genes and developmental events that shape the course of phenotypic development with a certain range of possible outcomes. Developmental change, not causal connections, is the focus" (pp. 155–156).

Mayr's (1976) formulation on this topic is also central to an interactional position: "The quality of the phenotype depends on the interaction of a very large number of genetic combinations with a large number of components of the environment" (p. 62). Such a view is emphasized by Cairns (1979a) in his evaluation of the research on the role of heredity and environment in individual differences in aggression. The differences obtained by selective breeding show a strong environmental specificity and can be modified by environmental social conditions to such an extent that the inherited differences can be eradicated.

Empirically, the interaction between genetic and environmental factors in the development of cognitive performance has been investigated by Fishbein (1979). In a longitudinal study using cross-sectional data she confirmed the hypothesis advanced by Scarr-Salapatek (1971) that the interactive relationship with regard to IQ differs between upper and lower class children. Plomin, DeFries and Loehlin (1977) examined the issues of interaction and correlation between genetic and environmental influences on human behavior that complicate behavior genetic analyses (cf. Plomin & Daniels, 1984). The process by which the individual develops within the boundaries and the potentialities offered by inherited properties has been discussed by Lerner (1984) in terms of plasticity of the organism.

The view adopted in the present research was formulated by Kuo (1967) when he summarized his propositions about *behavior potentials:*

> To put it more plainly, the science of behavior from the epigenetic point of view is not a "psychology without heredity", but it is a science based on the idea that heredity means merely the fact that the zygote starts to develop with an extremely wide (especially in higher vertebrates), but not unlimited, range of behavior potentials, only a very small fraction of which can be realized during its developmental history." (p. 128)

Given this perspective, the question of how much of a certain aspect of individual functioning is inherited becomes less meaningful. The formulation reflects the proposition by Anastasi (1958) that the appropriate question is not "how much" but rather "how" (cf. also Hofer, 1981).[+]

[+]An interesting position on the role of genetic and environmental factors is taken by sociobiologists, who claim that the evolution of the human species is directed by a process of reciprocal interaction between social and cultural factors on the one hand and genetic factors on the other (Wilson, 1975; see also Rushton, 1984). This view has been strongly criticized by Kitcher (1985) and Kamin (1985), among others.

DEVELOPMENT: A DYNAMIC PROCESS OF INTERACTION AMONG SYSTEMS

Structures and Processes

According to the interactional perspective, an individual develops and functions as an integrated psychological and biological being in a continuous reciprocal interaction with the environment. In this process of interaction, both the biological and psychological subsystems of the individual and the environment are in a process of transition into new states during the course of development. In the individual this occurs as a result of maturation and experience. The current interaction process is determined by the character and properties of existing mental and biological structures in the individual and by structures in the physical, cultural and social environment. For example, an individual's self-perception plays a strong role in the interpretation of stimuli and events in the environment at each specific stage of development. At the same time these new experiences contribute to change self-perceptions. This implies that the structures, both in the person and in the environment, change continuously, partially as a result of the processes. As Bell (1971) formulates the view with reference to parent-child interactions: "The basic principle underlying reciprocal influences in development arising from parent-offspring interaction is that of a moving bidirectional system in which the responses of each participant serve not only as the stimuli for the other but also change as a result of the same stimuli exchanges, leading to the possibility of extended response on the part of the other" (p. 822). This implies a spiral-like individual development in terms of person-environment interactions and interactions among the subsystems of an individual: it is continuously growing and changing (cf. Langer, 1969).

Levels of Structures and Processes

The interaction processes take place at various levels of the person-environment system, from the interaction between a cell and its environment in the early stage of development of the fetus up to the interaction between a person and the macroenvironment and between generations and their environments (Featherman & Lerner, 1985; Lerner, 1976a; Raush, 1977; Schneirla, 1972). The character of the interactional process will vary with the level of complexity in the hierarchy of the total system. For example, the character of the interaction process in which various neurotransmitters and cognitive factors are involved is different from the character of the interaction process between individuals and their environments as reflected in changes across generations (Elder, 1979) or in the evolution of the human species.

A Temporal Perspective

The formulations representing development as a spiral process that takes place with structures in continuous flux bring the temporal perspective of the interaction process into focus. This perspective will vary with the character of the system under consideration, in the sense that processes in systems at a lower level generally are characterized by a shorter time perspective than processes in systems at a higher level. For example, the interaction process between neurotransmitters takes place at a faster speed than the interaction process between generations and environments. (Cf. the discussion by Cairns & Cairns, 1985, and their distinction between short term interactions in the perspective of seconds and minutes and developmental interactions in the perspective of months and years.) This implies that the pace at which the structures in the individual and the environment change as a result of maturation, learning, and "experiences" varies with the character of the systems, especially the level of the subsystems. The anatomical structure of the fetus changes (as a result of the cell-environmental context interaction) at a much faster speed than the individual changes in the process of aging. Since systems at various levels are embedded in each other and are involved in a constant, dynamic, reciprocal interaction, the temporal perspective does not apply only to one subsystem at a time. The important implication of this view is that each subsystem must be analyzed in terms of its context in the total person-environment system and the manner in which it affects and is affected by other subsystems (cf. Öhman, 1986).

A certain process, for example leading to psychosomatic problems, can be studied in terms of what starts, directs, strengthens, and maintains it as well as in terms of its outcomes. Thus, as was emphasized in the previous section, an important characteristic of many processes is that the operating factors as well as the outcome(s) change across time, partly as a result of the process(es) in which they are involved. Over time the relative role of various factors may change. A certain factor, say a physiological or somatic factor, or a certain set of such factors, may start a process that is later strengthened, directed, and maintained mainly by other factors, for example cognitive-affective factors.

Implications for Developmental Research

The view that has been presented in this chapter has several implications for theory, planning, and interpretation of developmental research. The implications of direct interest for the planning and implementation of IDA will be discussed in Chapter 2. Here a few general comments will be made. The first comment is concerned with the effects of changes in the macrosystems in which individuals grow up. The possible confounding effects of age and cohort, if such changes are not considered and met appropriately by proper research designs, were discussed by Baltes and Schaie (1973), Baltes, Cornelius, and Nesselroade (1979), Schaie (1965) and Schaie and Baltes (1975), among others. These effects are obvious and must be taken into account. With reference to the importance of social relations, in which the family plays an essential role, an illustrative example can be found in the work of Bronfenbrenner (1958), who studied differences in parenting between lower-class and middle-class families. In some respects the class differences were reversed between 1930 and the 1950s. Such changes over time must be taken into account when investigating certain aspects of individual development. At the same time it should be emphasized that the extent to which the lawful principles that we are seeking are related to cohort properties varies with the kind of the problem and the character of the subsystem(s) of factors that are under consideration. For example, it is difficult to see how cohort effects could influence the lawfulness of the interaction process between cognitions, neurotransmitters and moods in individuals.

The second point is also concerned with the role of macrosystems for the individual development process. As noted earlier, the environment influences individual functioning in the person-environment interaction process at all levels, from the micro to the macrolevel. The environmental influence is thus not restricted to stimuli and events in the immediate, specific situations in which the individual appears. It may have profound importance for behavior in current situations as well as for the total life course of an individual, if, for example, he or she grows up in an urban instead of a rural area, in the mountains instead of the plains, in a Christian instead of a Hindu family. This circumstance naturally has to be considered when interpreting results from single studies performed in a certain culture under its particular conditions. It restricts to some extent the generalizations that can be made about individual development from single studies. At the same time it emphasizes the need for cross-cultural research on developmental issues in order to determine what is variant and what is invariant across cultures in individual development. Then additional, essential knowledge for the understanding of the lawful principles underlying individual development can be effectively distilled.

The third point deals with methodological implications. Psychological research in general has been characterized by methodological "monism," or even, in Koch's (1981) terminology, by methodological "fetishism." It should be clear from the foregoing discussion that the same methodology and research strategy cannot be applied without reservation to all levels of subsystems. There is no single "scientific" method that can be used for effective research on all types of problems. The specific character of the structures and processes for each subsystem and its relation to other subsystems makes an effective strategy and methodology for research on one subsystem and its interactions, but this is not always appropriate or effective for research on other subsystems of other characters and/or at other levels. The important point, that techniques most effective for description of the outcomes of developmental processes may not be effective for the analysis of the processes by which social patterns arise and are maintained or eliminated, was long ago made by Binet (see Cairns & Ornstein, 1979).

The final comment concerns the ability to attain the goal of psychology, as formulated in the introduction to this chapter. The comprehension and complexity of the phenomena to be considered in order to understand and explain why individuals think, feel, act, and react as they do in real life, makes the psychologist's task exceedingly complex. This has brought some researchers to a pessimistic view about the future of psychology as a science. In the past, Kant questioned whether psychology would ever reach the status of a science (Cofer, 1981). A similar pessimistic view has been expressed by more contemporary psychologists when discussing the complex interactions of factors that have to be taken into account in order to understand individual functioning (Cronbach, 1975; Gergen, 1973). It should be emphasized, however, that the litmus test of a scientific discipline can not be whether or not its phenomena are complex and hard to analyze: Had such a criterion been applied in the natural sciences from the beginning of their history, physics or chemistry would never have been regarded as scientific disciplines. The only criterion for a science is the appropriateness of its methods in dealing with relevant questions. Whenever processes display order and regularity on the basis of given structures, it is a scientific challenge to map this lawfulness of the order and the regularity (cf. Bateson, 1978).

The point of departure for an interactional perspective of individual functioning is that the life course of each individual takes place in a dynamic, reciprocal interaction process in which both the person and the environment change across time in a way that is characterized by order and regularity. There is, in principle, nothing mysterious or incomprehensible about the process, at least no more so than is valid for phenomena investigated by other scientific disciplines. Therefore, if research stays within the boundaries of natural limitations and does not extend the discipline to include existential problems as, for instance, why individuals have been given the opportunity to spend a short time in the universe, research should move forward. Of course, this does not imply that scientific methods will ever reveal the final truth about why individuals think, feel, act and react as they do; what can be done is to take steps in the direction of a better understanding of these phenomena. What is more important then than anything else is that we "use our brains" (Cronbach, 1975).

Chapter 2
THEORETICAL AND CONCEPTUAL CONSIDERATIONS

As emphasized in the preceding chapter, an interactional perspective has important implications for theory and empirical research on individual functioning, both in a current and in a developmental perspective. In this chapter conceptual and theoretical considerations that have been of central concern in the planning and implementation of IDA over the period of investigation from childhood to adulthood are discussed.

A PERSON APPROACH

Human functioning has a complex causal background. Even simple psychological phenomena can be influenced and caused by a multitude of factors within the individual and the environment. The operating factors in these processes underlying individual behavior change with age, partly as a result of the individual's developmental history. Psychological and biological factors in the individual as well as physical, social, and cultural factors in the environment interact with developmental levels. The effect of this process on the individual differs depending on the age and the course of the individual's earlier development (Emmerich, 1968; Flavell, 1971; Kagan, 1971; McCall, 1977; Moss & Susman, 1980; Wohlwill, 1980). As emphasized in Chapter 1, the implication of this general view is that the individual rather than the variable provides the central conceptual frame for the analysis of developmental processes.

The use of the term "holistic" to denote this position may contribute more to confusion than to clarification, because of its historical connotations. The same may be the case with the term "person approach." Thus, a few comments are required.

Of course, adoption of a holistic, person approach does not mean that we should not analyze and investigate those specific psychological and biological aspects of individual functionings which are usually discussed and investigated in terms of "variables." Such analyses are among the required fundaments for understanding the functioning of the totality. Mayr's formulation (1976) with respect to biology has relevance for psychology as well: "The past history of biology has shown that progress is equally inhibited by an antiintellectual holism and a purely atomistic reductionism" (p. 72). As emphasized earlier, the proper way to investigate the structures and processes varies with the level of complexity in the relevant hierarchy of systems. A person approach, then, does not inhibit analysis of specific aspects of individual functioning. Instead, it has two main affirmative implications: (a) the choice and theoretical analyses of specific aspects for investigation and of the measurement models used for data collection and data treatment should take place within the frame of the individual functioning as a totality and be guided by such a view and; (b) the results of empirical investigations of single aspects should be interpreted within the larger holistic frame of reference and integrated with results from other studies using the same frame.

One conclusion that can be drawn from earlier research is that truly effective analyses of development have begun with a holistic view of function, meaning, and adaptation and only afterward looked more precisely into specific mechanisms to find out how the relevant systems work. In the holistic person approach to the analysis of individual functioning it is not possible to make much headway if the analysis ignores the mechanisms of learning and biology and how they operate together at all levels of the relevant structures and processes. These ideas about mechanisms then have to be applied to the system as a whole.

Another important aspect of a holistic approach was well summarized by Weiss (1969):

> When people use the phrase "The whole is more than the sum of its parts," the term 'more' is often interpreted as an algebraic term referring to numbers. However, a living cell certainly does not have more content, mass or volume than is constituted by the aggregate mass of molecules which it comprises. As I have tried to illustrate in a recent article (P. W., 1967), the 'more' (than the sum of parts) in the above tenet does not at all refer to any measureable quantity in the observed systems themselves; it refers solely to the necessity for the observer to supplement the sum of statements that can be made about separate parts by any such additional statements as will be needed to describe the *collective behavior* of the parts, when in an organized group. In carrying out this upgrading process, he is in effect doing no more than *restoring information content* that has been lost on the way down in the progressive analysis of the unitary universe into abstracted elements. (p. 11)

A main conclusion of the above discussion is that there is no real contradiction between a person approach to developmental research and theoretical analyses and empirical investigations of specific aspects of structures and processes that are involved and specific mechanisms that are operating in the processes (cf. Goodfield, 1974; McCall, 1981). What is important to note is whether the object of interest is confined to the particular aspect per se or whether it is analyzed and investigated as one aspect of an integrated whole.

Generalizations in a Person Approach

Critics of a person-centered approach to psychological research acknowledge that it may be appropriate for case studies of individuals, but claim that it does not permit the same degree of generalizations as is possible in the variable-centered (often discussed in terms of a nomothetic) approach (cf. Nunnally, 1967). However, as should be clear from the discussion earlier in this book, a person-centered approach is not limited to being used only for case studies if appropriate methodology is used. Both in principle and in practice, generalizations can be made with the same degree of effectiveness when using this approach as when using the variable-centered approach. At the highest level of generality, individuals can be categorized into subgroups and sub-subgroups on the basis of their patterns of relevant individual characteristics, comparisons can be made across categories, and generalizations can be made to the appropriate population with respect to known characteristics.

Individual Differences

One of the main approaches to understanding the lawfulness of individual functioning in general is by the investigation of individual differences. Therefore, this becomes a central issue in developmental research. As Vale and Vale (1969) stated, ". . . there appears to be little opportunity for psychology to become a science of general laws without systematically including individual differences in the search, and general laws are the business of all of psychology." Similarly, Block (1982) concluded: "Continued ignoring of the presence and implications of individual differences is likely to continue to slow the progression of the sciences of psychology" (p. 294) (cf. also Eysenck, 1983; Underwood, 1975; Wachs, 1977).

To some extent, the potentials and limits with which individuals are born shape the character of their interaction with the environment. They determine the character of an individual's method of selecting external information and handling it in interactional processes with the environment, and they determine, to some extent, the environment's approach to dealing with the individual, that is, the experiences that an individual has. Again, it is in the interaction process that individuals develop, within the boundaries of their inherited predispositions, the cognitions and conceptions of the outer world that form the frame of reference for the aspects of the environment that they attend to, seek or avoid. In ss of interaction with the environment, individuals with inherited ..., as well as differences based on experiences with the environment, ...m methods of functioning in which differences are manifested in cognitive-affective factors (competence, goals, values, motives, strategies, etc.), temperament (impulsivity, sociability, activity, etc.), and conduct (aggressiveness, restlessness, Type A behavior, etc.) (cf. Scarr & McCartney, 1983). What characterizes and distinguishes a person from other persons is the patterning of these various aspects of the individual functioning as a totality. This implies, among other things, that we have to complement the traditional, variable-oriented methods for data treatment of individual differences with methods that describe individuals and permit the study of individual differences in terms of patterns of relevant variables. In Chapter 8 an empirical illustration of such an approach is presented. One advantage of a pattern-oriented approach as it is applied there is that generalizations refer to persons or groups of persons and that is what we are ultimately interested in.

For a discussion about appropriate methodological collection and treatment of data in terms of patterns, a distinction between *retro-similarity* and *pro-similarity* in patterns, suggested by Bergman and Magnusson (in preparation), may be helpful. Bergman and Magnusson assume that individuals with a similar patterning of factors at a later stage of development will show more similar patterns at an earlier stage of development than individuals with a similar pattern at an early age will show similar patterns at a later age. Thus, they assume higher retro-similarity than pro-similarity with respect to patterns of factors in the development process. From this it follows that prospective and retrospective analyses of the same multivariate longitudinal data will not necessarily lead to the same conclusions. The alternative methods of prospective and retrospective analyses of longitudinal data both contribute to the understanding of the developmental process, especially if used in conjunction.

THEORY AND OBSERVATION IN PSYCHOLOGICAL RESEARCH

In Chapter 1 the importance of a general theoretical frame of reference for developmental research and the influence of such a frame on the planning and implementation of the empirical part of the longitudinal project was emphasized. On a less abstract theoretical level, the question arises as to what extent the project has been directed by the investigation of hypotheses derived from a certain theory for development.

One interpretation of the role of theory vs. observation in psychological research, as it pertains to the design of our research, is as follows. One of the main causes of fragmentation of psychological theory and research (and thus one of the main reasons for the limited impact psychology has had even in fields of central interest for individuals and societies) is its use, or rather misuse, of theory in the total research process. In some fields it has long been a fashion and sometimes a necessity to start with a theory from which hypotheses are then derived and tested, as Cronbach (1975) noted. "Model building and hypothesis testing became the ruling ideal, and research problems were increasingly chosen to fit that mode" (p. 116). In this tradition the world of phenomena has been used to test the validity of theories rather than to identify central problem areas. In such a research process, technically sophisticated but irrelevant theories, subtheories, and fragments of theories often multiply and persist, yielding results of little importance for the solution of relevant problems and offering no knowledge of interest to anyone, except the researchers who defend or oppose the theory (cf. Fishman & Neigher, 1982; Koch, 1961; McNemar, 1960).

Examples of this tradition and the consequences for development are easy to find. For a long time learning theories presented by Hull, Guthrie, Tolman, and others formed the starting point for research in the area of learning. Many researchers spent decades, using well-equipped laboratories and large research grants, to test the relevance for human behavior of theories originally formulated to explain animal behavior. Seldom was a problem in the real world of education and schooling formulated and tested in the real world of education and schooling. The net result of all this testing of theory is now acknowledged to be very limited: little more than what good teachers have known and applied for centuries (cf. Koch, 1959). A second example can be taken from research on decision making during the 1960s and 1970s (cf. Cartwright, 1973). Any observation and analysis of the process underlying an individual's decisions show that emotions play a strong and sometimes decisive role in many, if not most, of the important decisions we make. The models most commonly used in theory-driven research on decision making excluded emotions from the models and attempted to control "extraneous" emotional intrusion in the experiments. Secondly, the experimental research ignored the social context of decision making. If we know anything about choices in actual life, it is that they are not made in a social vacuum. The mathematical models of the decision process, however, were divorced from the dynamics of the social ecology and the relations with other individuals, as well as from intraorganismic events of an emotional sort. These circumstances may be one reason why research and practice in decision making have little to do with each other, as summarized by Fishman and Peterson (in press). If research on decision making had started from direct observation and careful psychological and conceptual analyses of decision making in real life situations, the empirical research would have become more realistic and thereby more effective than it has been until recently.

The situation with respect to the role of theory in research discussed above has not only occurred in psychology. In an evaluation of the situation in economics, the Nobel laureate Leontief (1982) expressed his dissatisfaction with the dominating trend in this discipline: "Page after page of professional journals are filled with mathematical formulas leading the reader from sets of more or less plausible but entirely arbitrary assumptions to precisely stated but irrelevant theoretical conclusions" (p. 104). Leontief argued strongly for a research strategy in which theorizing is more closely bound to direct observations and analyses of the phenomena under consideration.

A starting point for the planning of the longitudinal project, with strong guiding effects on the choice of observables and of methods for observation and data treatment, has been the view that our problems should be rooted in the world of real life phenomena, not in the world of theory; "The key is to know the phenomenon" (Cairns, 1979b, p. 198). This view makes observation and analyses of phenomena the point of departure for the formulation of theories and for the identification of problems for investigation. The role of theory is to provide coherent explanations for observations of the phenomena under investigation.

Observations and descriptions of the phenomena under consideration have a long tradition in developmental research. Classical examples, which also show the importance of using careful observations as the basis for theorizing, can be found in Baldwin (1895, 1897), Binet (1909), Gesell (1928), Gesell and Thompson (1934) and Piaget (1928). Charlotte Bühler in the 1930s demanded that her graduate students devote one semester to the careful observation and description of a child, without interpretation and evaluation of what they observed (Hans Kreitler, personal communication). Also during the period when theoretical hypothesis testing dominated psychological research, researchers in development maintained the tradition of direct observation (see e.g., Ainsworth, 1983; Bowlby, 1951; Radke-Yarrow & Kuczynski, 1983). An excellent example of the usefulness of direct observation in natural settings as a basis for theory-building on important developmental issues was presented by Raush and his collegues in a series of articles on their investigations of aggressiveness among boys using an interaction approach (Raush, 1965; Raush, Dittmann, & Taylor, 1959; Raush, Farbman, & Llewellyn, 1960).

A possible explanation for the misuse of theory in much psychological work can be traced to a misunderstanding of the role of theory in natural sciences upon which many researchers in psychology have based their ideals (Cronbach, 1975). Actually, close study of the natural sciences demonstrates the importance of observation and description as the basis for solid scientific progress. The most striking examples can be found in biology, botany, and ethology, where careful direct observation and description of the phenomena form the basic elements of the research process. In physics, chemistry, and astronomy, observations of deviations from expectations are important contributions to the development of new and more productive theories.

Of course, the foregoing points do not imply that theory is meaningless or useless. Scientific research can never be restricted to a collection of uninterpreted observations. The point is how theory is used. Theory has no value in itself. Its value depends on the extent to which it helps to explain the phenomena and the real world processes under investigation.

The direct implication of this view for our project has been that it was not planned to test a specific theory about individual development in a specific area of individual functioning. This view, of course, has not excluded the use of theory. The planning of the project, the choice of observables and methods for data collection and treatment were based on careful analyses of available knowledge and existing theories. These analyses were directed mainly toward identifying aspects important to observe at various stages of development. And, of course, the process of data treatment and interpretation of the empirical results has been guided by the general theoretical framework presented in this chapter and in Chapters 1 and 3. With careful observations and descriptions of phenomena, we hope to contribute to the foundation of more effective theories of development.

PREDICTION VS. EXPLANATION AND UNDERSTANDING OF LAWFULNESS

The goal of scientific psychology is to understand and explain why individuals think, feel, act, and react as they do in real life situations. This formulation has a direct impact on the choice between two of the contrasting ultimate goals for scientific psychology, namely, prediction of individual functioning and explanation and understanding of the lawfulness of individual functioning. The term "lawfulness" is used here to describe behavior that proceeds according to a set of identifiable principles, but is not necessarily predictable.

Prediction as the main goal of theory and empirical research in psychology was explicitly and distinctively formulated by J. B. Watson (1913) in the opening sentences of his article "Psychology as the Behaviorist Views It": "Psychology as the behaviorist views it is a purely objective natural science. Its theoretical goal is the prediction and control of behavior" (p. 158). The view expressed by Watson and followers has been very influential. Even in those areas in which a pure behavioristic view has not dominated theory and empirical research, perfect prediction has become the ultimate goal, and successful prediction the main criterion for the scientific status of psychology.

The discussion about prediction vs. explanation of lawfulness is not just a theoretical matter. The concept of prediction as a goal for psychology is closely connected with a mechanistic model of man (Overton & Reese, 1973). Associated with this view is the concept of cause and effect, independent and dependent variables, as well as analyses of predictor and criterion relations. Psychological research in which cause and effect are studied as a unidirectional relationship between independent and dependent variables, and between predictors and criteria, is often motivated by the goal of accurate prediction. This view is also reflected in the operationalization of psychological concepts, the choice of variables to be studied at different age levels in developmental research, the methods for treatment of data, and the interpretation of empirical results.

Two main aspects of an interactional model of the human have direct implications for the discussion regarding prediction vs. lawfulness as the main goal of scientific psychology. First, the main object of interest is the individual as a totality, in which each aspect of the individual derives functional meaning from its role in the total functioning of the individual. Second, the individual functions and develops in a dynamic, continuous interaction with his environment. Thus, the primary interest is in the lawfulness and continuity of the processes involved in the individual's functioning in current situations and in the structures and processes that are involved in the changes in the individual's way of functioning across time due to maturation and experience in a developmental perspective. Given the complex, often non-linear interplay of biological and mental subsystems of factors within the individual and the complex interplay between the individual and the environment that is operating in relation to the individual in a probabilistic, sometimes very uncertain and unpredictable way, it is unrealistic to hope for high prediction of molar, social behavior from one situation to another or over age spans (Bandura, 1982; Scarr, 1981). To foresee molar social behavior is as difficult for psychologists as it is for meteorologists to foresee the weather from day to day.

The problems that a researcher in psychology encounters resemble those encountered by a researcher in the area of meteorology. Weather and climate are best described by process models that have some characteristics in common with psychological models; many factors are involved, they operate in a continuous and bi-directional interaction, they interplay in a very complex, non-linear way, etc. Though a great deal is known about the factors that are associated with changes in the weather and the way they operate in relation to each other, meteorologists cannot predict change in the weather from day to day with any great degree of certainty. However, this has not diminished the status of meteorology in the scientific community.

It is sometimes argued that high predictability is one prerequisite for real scientific explanation. Scriven (1959) has countered this view in his discussion of explanation and prediction in evolutionary theory: "Satisfactory explanation of the past is possible even when prediction of the future is impossible" (p. 477). Later he states ". . . the thesis of this article is that scientific explanation is perfectly possible in the irregular subjects even when prediction is precluded. One consequence of this view is that the impossibility of a Newtonian revolution in the social sciences, a position that I would maintain on other grounds, is not fatal to their status as sciences" (p. 477).[1]

Of course, prediction is a useful conceptual and methodological tool in planning, implementing, and interpreting certain kinds of empirical research,

---

[1]The scientific status of psychoanalysis as a model for personality is sometimes questioned on the grounds that it cannot predict behavior. According to the standpoint formulated here this argument is not valid, although the scientific status of psychoanalysis may be questioned on other grounds.

for example, testing hypotheses derived from explicitly formulated theories and models of individual functioning at a certain level of analysis. In the application of psychological knowledge as a basis for personnel selection, for decision making in planning, and in numerous other practical situations, the level of certainty of the prediction made using such knowledge is of basic interest, as it is for a meteorologist and for those who are dependent on his weather forecasts. These are reasons for both meteorologists and psychologists to determine the bounds of certainty within which their predictions are asserted. What is argued against is the tendency to make high prediction the primary goal of scientific psychology, independent of the level and type of analysis of human functioning. Pursuit of such a goal is not only futile, given the character of the processes under examination; it is destructive and hampers real progress in scientific psychology.

CONSISTENCY, STABILITY, AND CHANGE IN DEVELOPMENT

A central issue for developmental theory and empirical research is that of continuity and change, with respect to specific aspects as well as to the total functioning of an individual in the developmental process. This issue raises conceptual, theoretical, and methodological problems that have implications for developmental research strategies. Central problems have been discussed by Bell, Weller, and Waldrop (1971), Block (1982), Brim and Kagan (1980), Cairns (1979a), Cairns and Hood (1983), Emmerich (1964; 1968), Flavell (1971; 1982), Kagan (1971), Lerner (1984), Sroufe (1979), Sroufe and Rutter (1984), and Wohlwill (1973), among others. The theoretical discussion on all these problems will not be reviewed here. Only a comment on cross situational vs. temporal stability will be made before summarizing the view on stability and change that has formed the basis for IDA's research strategy.

It must be kept in mind that the problems with appropriate models for cross-situational consistency and developmental consistency in individual functioning are orthogonal to each other. Data for the elucidation of one of these types of models cannot be used for the elucidation of the other, except under certain conditions and for specific purposes, which must be specified in each case.

There are two competing models for the explanation of cross-situational consistency in individual functioning in terms of manifest behavior. The traditional "relative stability" model assumes that personality consistency is revealed in cross-situationally stable rank-orders of individuals for the behavior under consideration. The second model, the interactional model, assumes that consistency in behavior is reflected in coherent, partially *specific patterns* of stable and changing behaviors across situations, that characterize each individual (see Magnusson, 1976). It has been argued that the interactional model for cross-situational consistency has been weakened by coefficients showing high stability over time for specific aspects of behavior, for example, aggression, intelligence, and creativity (Magnusson & Backteman, 1978; Olweus, 1979). However, such coefficients for single aspects of behavior are compatible with both the traditional, relative stability model, and the interactional, pattern model for cross-situational consistency and do not discriminate between them.

Within the theoretical framework that was presented in Chapter 1, the general view on developmental continuity and change that has shaped the planning of the longitudinal project can be summarized as follows.

1. At each stage of development an individual functions as a totality. Development does not take place in single factors or aspects of functioning per se in isolation from the totality; it is the total organism that functions and develops.

2. As a result of maturation and experience in the process of interaction with the environment and among subsystems of mental and biological factors, individuals change their method of functioning across time. In the process of transition into new stages, the character of specific aspects as well as the role they play in the total pattern of operating factors may change. For example, the mental processes that are discussed and investigated as aspects of intelligence can be qualitatively different for varying ages and may change the total functioning of the individual. As shown first for the sea urchin egg and later also for the vertebrate embryo, some neurotransmitters, such as serotonin and dopamine, play one role in the developmental process and another role in current behavior (Gustafsson & Toneby, 1971; Lander, Wallace, & Krebs, 1981).

3. The developmental process and the changes that take place in the individual from one state to the other across time are coherent. Each state is lawfully related to the preceding one. Everything that happens does so within an organism that has a history of mental and biological processes. This does not mean that one state can be predicted from an earlier one or that these need to be "closely dependent relations between all the successive stages" (Kagan, 1978, p. 72), only that development is coherent and lawful.

As Cairns and Cairns (1985) have emphasized, early theorists using very different approaches have assumed (mostly implicitly), that development leads to increasingly higher levels of organismic adaptation and competence and that development is synonymous with progress. When this assumption is examined in the light of a life span perspective, it loses credibility (Kagan, 1983; Pollack, 1983). As pointed out by Sroufe and Rutter (1984), even regression is a part of a lawful, coherent process of development that can be explained in terms of the individual's earlier life history and current environmental factors.

4. One characteristic of the developmental process is the natural change that occurs in all normal development. For example, normal development includes a growth spurt in height that generally occurs two years earlier for girls than for boys within a given culture (cf. the discussion about readiness, Rasmusson, 1983). At the same time there are inherited, biological individual differences, with respect to potentialities and limits in competence and vulnerability, which are reflected in differences in the individual's interaction with the environment. These inherited dispositions form the basis and the limits for change and stability.

5. An important characteristic of the developmental process is the individual differences that occur with respect to the pace and patterning of factors in the development process (cf. Bateson, 1978; Hinde & Bateson, 1984; Uzgiris, 1977), implying that important individual differences in development can be investigated best in terms of patterns of operating factors, as was emphasized earlier.

DISPOSITIONS AND "TRAITS" IN AN INTERACTIONAL MODEL OF MAN

A common reaction to the interactional approach to psychological research requires comment, as it has been influential in the debate, even though it is based on a misunderstanding of the interactional position. This view has been expressed by Plomin (1986) in a discussion about models for the interplay of genetic and environmental factors in the developmental process: "If interactionism were to be believed, it would imply that "main effects" cannot be found because everything interacts with everything else" (Plomin, 1986, p. 249). Two comments are pertinent. First, the description of a multiple determined process as characterized by continuous interaction does not exclude the investigation of main factors operating in this process. Second, and particularly important for the present discussion, is the formulation by Plomin that reflects the idea that an interactional position excludes the identification of enduring personality dispositions (traits) as main factors. (The concept of trait is often ambiguously defined.) Empirical studies demonstrating the existence and important role of such dispositions have commonly been used as arguments against an interactional position. Such arguments motivate a few comments.

As briefly described and discussed earlier, two contrasting, general models for cross-situational consistency of individual functioning can be identified. The difference between the two models is in their treatment of differential situational effects on behavior. The traditional relative stability model assumes that individual differences can be interpreted in terms of dispositions that are manifested consistently across situations. This is expressed in data in stable rank orders of individuals for each specific aspect of individual functioning that is under consideration (Figure 2.1a). By contrast, an interactional model for cross-situational consistency assumes that individual differences are reflected in partially specific cross-situational patterns of individual functioning, reflected in data for a certain aspect in partially specific cross-situational profiles (Figure 2.1b). Each individual profile is then defined by two parameters, level and form. Individual differences with respect to level make up the variance due to persons as the main effect in a person-situation matrix of data for a specific type of behavior. Individual profile differences remaining when the main effect due to individuals is screened out make up the variance due to interaction between individuals and

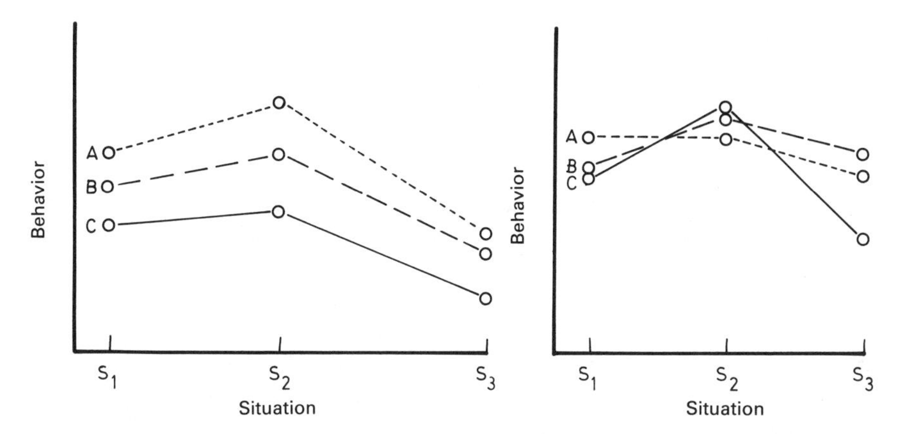

FIGURE 2.1. Cross-situational profiles for individuals A, B, and C for a certain aspect of behavior according to (a) the relative stability model (left) and (b) the interactionistic model (right).

the situational conditions under which the observations have been made if all variance is assumed to be reliable.

Therefore, an interactional model for cross-situational consistency in a current perspective assumes the existence of consistent personality dispositions which direct behavior and thereby influence and determine, to some extent, the person's interaction with the environment. The existence of such dispositions is an everyday experience and provides the basis for purposeful interaction with the environment and effective intersocial relationships. Temperamental features change during the course of development and the extent to which this is the case is dependent on maturational factors (Achenbach, 1985) as well as on the individual's experience of environmental factors (Gottlieb, 1976a,b; 1981). Individual differences in this respect are due to differences in inherited dispositions and differences in the extent to which the individuals stay in the same homogeneous environment or range through more heterogeneous environments.

The important role of stable features of an individual's way of functioning in a current perspective has been amply demonstrated in personality research. Individual differences in source and surface traits (Cattell, 1965), in sociability (Guilford 1959), in impulsivity, neuroticism, and introversion (Eysenck, 1967), and in sensation-seeking (Zuckerman, 1979) influence to some extent the type of situations that individuals seek and avoid and how they function once they are in a certain situation (Stagner, 1977). In a developmental perspective, Block and Block (1980) have presented empirical support for the existence of stable individual differences from 3½ to 7½ years, in both laboratory and observational situations, with respect to "Ego control" (the degree of control the child maintains over impulses, wishes, and desires) and "Ego resiliency" (flexibility of controls). Buss and Plomin (1984) have emphasized the role of stable temperament variables (emotionality, activity, and sociability), as have Lerner and Lerner (1983). The existence of important, relatively enduring dispositions that characterize a person's way of functioning in a current as well as in a longitudinal perspective is central to an interactional view.

Individual differences with respect to enduring dispositions can be genetically determined and only slightly influenced by environmental factors while others can develop from the process of interaction with the cultural environment (Buss & Plomin, 1984; Lerner, 1984; Loehlin, 1982). To a certain extent, the character of both the interaction among psychological and biological subsystems and that between the individual and the environment is dependent on genetically determined dispositions. The whole developmental process takes place within the frame that genetically determined dispositions impose for the interactional processes.

The interactional approach to individual functioning in a current perspective differs from the traditional trait oriented approach in an important respect connected with the appropriate measurement model. In traditional measurement models, observed individual differences in cross-situational profiles have been regarded as errors and wiped out. Frequently this is done implicitly by using non-situation-specific data, as in the case with data from traditional inventories, without taking into consideration the situation-specific adaptation of the individuals. In contrast, as emphasized by Magnusson (1976), an interactional view regards cross-situational differences in adaptation to situation specific characteristics as valuable and valid information that can be used beyond the information about cross-situationally valid, general individual differences, with respect to level of adaptation to various situations (cf. also Mischel & Peake, 1982).

To summarize, an important element of an interactional view is the assumption of the existence of enduring individual dispositions. Thus there is no contradiction on this issue between an interactional position and that of traditional trait theory. This position is also basic for behavioral genetics. The main distinction to be made is that an interactional position claims that it is not possible to understand and explain why individuals function as they do in specific situations solely in terms of single traits or combinations of single traits as they are usually operationally defined in personality and developmental research (cf. Diener & Larsen, 1984). This issue will be dealt with further in Chapter 3 when the importance of matching the type of data to the level of structures and processes under consideration is discussed.

MALADJUSTMENT IN THE INTERACTIONAL MODEL

An essential issue for developmental research is the conditions under which the interactional process leads to maladjustment in terms of (a) deviations from a *biological* norm of physical health, as in psychosomatic diseases, (b) deviations from a *psychological* norm of subjective satisfaction, as in depression, and/or (c) deviation from a *social* norm, as in criminality (cf. Wenar, 1982). An interactional view implies that maladaptive functioning in these respects is not determined by either person or environmental factors, whether in a current or in a developmental perspective. Rather, maladaptive behavior is seen as the joint outcome of *both* person and environment factors.

For a particular individual, the kind and degree of current maladaptive functioning is determined by the relation between the individual's psychological-biological vulnerability and the kind and intensity of environmental provocation to which he or she is subjected. From a developmental point of view, both psychological and biological vulnerability can be restricted to specific types of reactions (skin or heart reactions) or be general enough to include the whole organism. Provocations in the environment can be restricted to specific situations for a person (phobic situations, for example) or include almost all types of environments. The effective environmental factors may be physical, social, cultural, or psychological and may operate independently or in conjunction with each other. Individuals can be born with specific or general vulnerability. Thus, there are biological individual differences in vulnerability from the beginning. For example, some empirical studies suggest that certain developmental gender differences in vulnerability exist (Bayley, 1966; Bergman, 1981; Hutt, 1972; Kamin, 1978).

The general idea that individual malfunctioning can be affected by both environmental conditions and person dispositions is not new. For example, the Swedish psychiatrist Sjöbring, whose work during the first part of this century has been met with renewed interest in recent years, emphasized that many diseases were the outcome of an interplay among hereditary biological factors, physical and social aspects of the environment and the individual's experiences in a wide sense (Sjöbring, 1958). Sjöbring's interactional view is reflected in the following formulation: "What characterizes a person is the way in which he/she relates to a certain object, a certain situation, fulfilling a certain purpose" (p. 16). In the area of stress, Appley and Trumbull (1967) formulated this view that incorporates a developmental perspective:

> It is perhaps doubtful that there is such a thing as a general stress-tolerance in people. There is more likely to be a greater or lesser insulation from the effects of certain kinds of stress-producers rather than of stress. . . . It seems more likely that there are differing thresholds, depending upon the kinds of threats that are encountered and that individuals must be differentially vulnerable to different kinds of stressors. . . . To know what conditions of the environment are likely to be effective for the particular person, the motivational structure and prior history of the individual would have to be taken into account. (p. 11)

The general view on maladaptation reflected here and in formulations by other researchers has been applied by Depue and his coworkers in their discussion about depression from a psychobiological perspective (Depue, Monroe, & Shackman, 1979), by Brown and Harris (1978a, b; 1980) in their discussion of a vulnerability model for the origin of depression, by Weiner (1977) in his analysis of the background to psychosomatic diseases, by Zubin and coworkers (Zubin & Spring, 1977; Zubin & Steinhauer, 1981) in their work on schizophrenia, and by Kagan (1983) in his review of research on stress and coping in early development. A recent review of psychopathology in an interactional perspective by Öhman and Magnusson (1987) summarizes much of this work.

As reflected in the foregoing citation, Sjöbring emphasized the role of an individual's experience in the outbreak of physical and mental disease. This line of thought has been broadened and refined by the rapidly growing research on physiological processes in general, particularly on the role of neurotransmitters and hormones and their interaction with mental factors. Among other things this research has shown the importance of the environment in terms of the direct mechanistic influence of negative environmental physical conditions, but above all, in terms of how these conditions are perceived and interpreted by the people who are brought up and live in them. An individual's interpretation of a situation as threatening will in some way evoke physiological processes, which, if they occur frequently, will in the long run affect important systems, such as the immune system, with consequent effects on physical health (e.g., Jemmott & Locke, 1984; Maier & Landenslager, 1985; Stein, 1985; Stein & Schleifer, 1985). During recent years the role of social factors in this process, particularly the behavior of others and its interpretation by the individual, has been emphasized and led to the delineation of a specific field of research, *sociophysiology* (Barchas, 1984; Waid, 1984). These models again underline the importance of considering both psychological and biological factors in the person and environmental factors if the processes underlying mental and physical maladjustment are to be properly understood.

Patterns of Maladjustment

The patterning of aspects of individual functioning as characteristic of individuals and the importance of studying interindividual differences in terms of such patterns were emphasized earlier. This emphasis becomes particularly important when the object of interest is maladaptation or maladjustment, as it is in the longitudinal study. As reviewed by Rutter (1982), among others, the importance of a certain aspect of maladjustment in the developmental process depends on its role in a broader pattern of aspects (see also Meyer-Probst, Rösler, & Teichman, 1983). From an interactional standpoint, which emphasizes the study of individuals as totalities, it is natural to view maladjustment, intrinsic and/or extrinsic, as a syndrome of individual characteristics in which several symptoms are normally involved. How these symptoms are grouped together in individual patterns of maladjustment is a main subject of interest in IDA (see

Bergman, 1985; Bergman & Magnusson, 1983; Magnusson, 1985b; Magnusson & Bergman, 1984; Magnusson & Bergman, in press). The view that a certain type of maladjustment is characterized by a particular constellation of symptoms, "critical configurations," implies, among other things, that one single symptom, even in extreme cases, may not be an indicator of maladjustment. It is in combination with other symptoms that it becomes an indicator or maladjustment as we define it. This makes the study of critical configurations of relevant person factors in development a central task for longitudinal research on maladjustment.

The limitations of considering only single factors in the study of maladjustment are also obvious for environmental factors operating in the person-environment interaction process underlying maladjustment. Simple factor oriented research has led to the distinction of so called "risk factors." However, a low socioeconomic standard of the home of a child does not necessarily in itself predispose the child for later maladjustment. Most children raised under such conditions do not develop symptoms of maladjustment. A low socioeconomic standard may increase the probability that weak social norms and/or bad social relations among the family members will lead to maladjustment. This interaction between socioeconomic factors in the family, social relations among the family members, and extrinsic maladjustment in youngsters was empirically illustrated in a study in the project presented by Dunér and Magnusson (1979).

Extrinsic and Intrinsic Adjustment

The central issue of IDA is individual development. The development of an individual, from childhood to adulthood, takes place in a process characterized by continuous interaction between the individual and his/her environment. This process can be conceived of and investigated as a continuous process of adaptation and adjustment (Cairns, 1979a; Gould & Vrba, 1982; Prechtl, 1976; Sroufe, 1979b). For the planning of the project, particularly with respect to the choice of aspects of individual functioning to be examined, the view of development as a continuous adaptation process formed an essential framework. A basic distinction was then made between *extrinsic* and *intrinsic* adjustment.

Extrinsic adjustment is reflected in the agreement between the individual's behavior and the demands from the environment in two main respects, namely, (a) of norms for conduct and (b) levels of achievement. These demands define the individual's role in the social system. Intrinsic adjustment is reflected in the individual's own satisfaction with the situation. It can be regarded as the result of the agreement between the individual's own needs, values, and motives on the one hand, and the rewards received from those actions and the environment, on the other hand (Magnusson, et al., 1975).

The concepts of intrinsic and extrinsic adjustment have vague boundaries, but they offer the advantage of being related to the common distinction between behavior problems vs. personality problems in psychiatry and acting out vs. neuroticism in psychodynamic terminology. For teenagers, the interplay between these two aspects, i.e., the self-evaluation and the social environment's evaluation of them and the relation between their own behavior and their own values, norms, and motives, can be regarded as the hub around which much of the adolescent phase of development revolves. The development of this process plays an important role in the formation of a stable self-identity in which the individual's inner experiences and behavior are in harmony with the expectations, demands, and potentialities of the society.

PROTECTIVE AND MODIFYING FACTORS IN THE ADJUSTMENT PROCESS

The view presented previously on prediction versus explanation has consequences for the empirical investigation of the developmental background of adult maladjustment.

During the last few decades much research has been done on adult criminal behavior, alcoholism, and mental disease in terms of environmental factors in the upbringing conditions and/or in the current environments. Various aspects of so-called "risky" environments have been studied, and correlations for almost every aspect of upbringing conditions and adult maladjustment have been identified. This has been part of the strong environmentalistic movement in the behavioral sciences. At the same time, empirical research has demonstrated a significant relationship between biological parents and their offspring with respect to various maladaptive behaviors, even when the offspring have been adopted at an early age. These results indicate that individual genetic differences form a basis for differences in the development and outbreak of maladaptive behaviors (Bohman, 1978; Goodwin, Schulzinger, Moller, Hermansen, Winokur, & Guze, 1974; Mednick, Moffitt, Pollock, Talovic, Gabrielli, & Van Dusen, 1983).

Though the influence of environmental and genetic factors on maladaptive behaviors seems to be well established by empirical research, it should be emphasized that most of the relationships are weak. Indications suggest that a rather small group of subjects, with multiple maladaptive symptoms, account for the low level but consistently observed relationships between various aspects of the environment and adult maladaptation. This question is discussed and empirically illuminated in Chapter 8. As a matter of fact, most individuals who have been raised and/or live under "high risk" conditions have *not* become criminals or alcoholics and have *not* developed symptoms of mental disease. Some of those who have contributed most to society, through art, literature, science, politics, and industry, have come from such environments. One may wonder if it was not the very challenge of overcoming strong obstacles and hindrances that prepared them for their contributions (Magnusson, 1985c; Werner & Smith, 1982). In this perspective, one of the most interesting and important questions for research on individual development is why most of those who are raised under conditions characterized as high risk, develop into well-adjusted adults (Garmezy, 1976, 1983; Hartup, 1979; Rathjan & Foreyt, 1980; Rutter, 1979; Suomi, 1979).

From its beginning, the longitudinal project has devoted special attention to the role of protective factors that may help to modify the effects of negative environments in the socialization process. Special interest has been paid to critical configurations of relevant aspects that have positive implications for functioning.

Chapter 3

METHODOLOGICAL AND RESEARCH STRATEGICAL CONSIDERATIONS OF AN INTERACTIONAL PERSPECTIVE

INTRODUCTION

This chapter deals with methodological and strategical considerations for developmental research within the frame of an interactional perspective, with particular reference to issues of importance for the planning and interpretation of the empirical part of IDA. It is concerned with five main issues: (a) the importance of using data suited to the level of complexity of the structures and processes under consideration; (b) methodological and strategical problems connected with the existence of individual differences in growth curves for single variables; (c) models and methods for studying individual differences in development in terms of patterns; (d) the implications of a prospective approach to empirical developmental research; (e) methods of observation. These main topics also involve a discussion of the problems of aggregation of data, measurement of change, and retrospective versus prospective approaches to developmental research.

LEVELS OF ANALYSES

An important point of departure for a discussion of central methodological topics is that the data appropriate for the elucidation of structures and processes at one level of complexity, can yield meaningless results if applied to problems at other levels. Consequently, methods and procedures for data collection, such as the traditional type of inventories, that yield valuable data for the investigation of one type of problem at a certain level of complexity, may be useless and yield misleading data when applied to problems at another level. Ajzen and Fishbein

(1977) presented empirical support for this statement in their studies of behavior-attitude correlations. They showed that a prerequisite for a high correlation was that data refer to targets and actions at corresponding levels of specificity (cf. Sjöberg, 1982). The same point is valid for methods and procedures concerning data treatment (cf. Magnusson, 1987, and Chapter 7 in this book for an illustrative example). Therefore, a primary requirement for effective research on developmental problems is that the instruments and procedures for data collection and treatment, as well as the data per se, be appropriate to the character of the problem under consideration, with particular respect to the level of complexity of the structures and processes. The neglect of this critical issue is a cause of much confusion and misunderstanding in the fields of personality and development study. Since this distinction has important methodological implications for the choice of methods of data collection and data treatment and for the interpretation of the results in the longitudinal project, it is dealt with at some length in the following section.

Levels of Processes and Levels of Data

The preceding formulations stress that a necessary condition for any meaningful interpretation of empirical data as a basis for description and explanation of continuity and lawfulness is that researchers must be aware of and make explicit the level of complexity and generality at which they are working, with respect to the structure(s) and process(es) under consideration. In order to make the point clear, the weather can again serve as a useful analogy.

The appropriate data for studying interseasonal variations in temperature are aggregate data for seasons, obtained on the basis of observations all days of the year or on the basis of observations of a representative sampling of days for each season. Means for seasonal temperatures will be consistent and show, among other things, that most northern regions of the globe have higher temperature in the summer than in the winter. The seasonal curves will reflect, in descriptive terms, perfect consistency from one year to the other. This implies that there is perfect predictability concerning the difference in mean temperature between summer and winter. The descriptive lawfulness in seasonal variations can be explained in relation to the amount of sunshine, among other factors.

If we then direct our interest to the varieties in temperature at a lower level of complexity of the process, the variation from day to day, we will find that the descriptive curves for temperature from day to day show very little conspicuous lawfulness. And predictability, particularly of change in the weather, is surprisingly low, despite the high standard of meteorological research.

In order to understand and explain day-to-day changes in temperature the researcher has to analyze the processes, define all relevant factors, and plan and implement more finely graded analyses of the dynamic, reciprocal process of interaction among factors that contribute to determine changes in the daily temperature. (In this case a meteorologist is in a more favorable position than a psychologist, since he can at least sometimes simulate conditions in the laboratory, without too much distortion of the real conditions under which the temperature changes.)

Aggregate data across days over one season can be regarded as measures of temperature as a "trait," a stable characteristic of the weather. Nevertheless, such aggregate data (even aggregations of observations across days within a week) do not help to improve day-to-day prediction of changes in the temperature or to demonstrate any lawfulness in the changes from day to day. Rather, if the two levels of analysis are not clearly distinguished but are mixed in the interpretation, the picture can be described (as it is in many reviews on important issues in psychology) as "confusing," since the temperature on a certain day in the winter may be warmer than on a certain day in the summer. This example has been presented in detail in order to demonstrate as clearly as possible one misunderstanding about the central problem in the discussion of personality consistency and person-situation interaction and the misuse of results from empirical research as a basis for various interpretations of that issue.

Two points should be clear from the analogy of the weather. First, the level of the process that is under consideration must be clearly distinguished. Second, the empirical data must be appropriate to the level of the process at which the problem is defined. The debate on personality consistency becomes meaningless and fruitless when these two criteria are not met, that is, when consistency in aggregate data at the trait level is used as an argument against the existence of person-situation interactions and when individual differences in cross-situational profiles, based on situation-specific data, are used as an argument against the existence of personality traits. One well-known attempt to elucidate the problem of personality consistency, presented by Epstein (1979, 1980), may further illustrate these points.

Epstein obtained his empirical data by observation of a certain behavior for a sample of individuals over 28 consecutive days. He found that the coefficients for the correlation between observations for single days were very low, indicating low relative stability in the type of behavior he had observed. He also found that by aggregating data for days with even numbers and data for days with odd numbers in successively greater numbers, the coefficients for the correlation between data aggregated for days with odd and even numbers increased gradually. This is what would be the predicted outcome of calculations of coefficients between measures of the temperature for odd and even days, respectively, during a certain season. What do these results actually demonstrate? First, they reflect the fact that there is low relative stability from day to day in the type of behavior observed. One comment on this observation has been that behavior is unreliable. This statement is confusing and based on a misunderstanding of the concept of reliability, as a psychometric, precisely defined concept. One example of the confusion is the discussion presented by Lamiell (1982). Reliability is a characteristic of the method for observation and a property of the data, not a property of the persons being studied or of observables. Behavior at each specific moment of the behavioral process is exactly what it is, and it can be observed and measured with varying degrees of reliability. The low coefficients for the correlation between observations for single days clearly show that the lawful consistency of behavior from one situation to another is not reflected in stable rank orders of individuals from situation to situation and cannot be sufficiently explained solely in terms of enduring personality dispositions.

Second, by increasing coefficients for the correlation between successively more aggregated observations across days, Epstein has empirically demonstrated the existence of rather enduring personality dispositions. To some extent, such dispositions determine the types of situations that are sought and what types are avoided as well as individual functioning in certain types of situations, as discussed in Chapter 2. Thus, he has demonstrated something that no serious interactionist has argued against. Rather, an interactional model assumes that such dispositions play an important role in an individual's total psychological and biological processes, not least in person-environment interactions.

It should be clear from the foregoing discussion that data aggregated across situations cannot be used for descriptive or explanatory purposes when cross-situational variation in individual functioning is the problem under consideration. The same conclusion about cross-situational variation can be drawn from the results of empirical studies demonstrating low rank-order stability across situations for specific behaviors (Bowers, 1973; Magnusson, 1976).

Epstein's study and the ensuing discussion of his results clearly demonstrate the necessity to meet the two criteria discussed earlier for any meaningful discussion about consistency and continuity in the processes underlying individual functioning, both in the current and in the developmental perspectives. When researchers are not clear with respect to the level of the processes at which they are working and the necessary linkage between the level of the processes and the appropriate type of data, both the discussion of the results of single studies and the discussion in the entire field becomes confusing. The real fact is that what at the surface seem to be inconsistencies in empirical results often contain valid and valuable information if they are obtained in carefully planned studies and are interpreted in the appropriate perspective.

The weather analogy can also be used to draw conclusions that have serious implications for research strategy in psychology. For a full understanding of the total process of individual functioning it is necessary to work at all levels of the processes, from the continuous, reciprocal person-environment interaction process to the process of interaction between genes and their context. Again, it is important that for each specific study, the level of analysis is made clear and that methods used for observation yield appropriate data. Methods for data treatment that are consistent with the level of the process under consideration must also be used. For IDA this has meant employing all available forms of observation;

inventories, tests, ratings, laboratory tests, direct observations in natural situations, and so forth. For treatment of data all relevant methods, analysis of variance and other forms of regression analysis, factor analysis, cluster analysis, etc., have been applied. Situation-specific data and aggregated data (at various levels of aggregation) have been used, depending on the level of the problem.

Aggregation of Observations Across Situations

The preceding discussion raises the issue regarding aggregation of data across situations in order to obtain reliable and valid measures of individual functioning at the trait level (for a comprehensive discussion of this issue the reader is referred to Blalock, 1982). Aggregation of observations across situations implies that situation-specific information is lost and consequently that the analyses using such data will concern structures and processes at a nonsituation-specific level. A datum representing this general level is *nonsituation-specific*. It is obtained as an aggregated sum or mean of observations across situations of various types. Thus nonsituation-specific data are not representative of, nor do they reflect, individual responses in a specific situation. Data of this type are the most common in traditional personality research within the traditional measurement model.

A nonsituation-specific datum can be obtained by aggregation in two main ways. The first type of aggregation is in terms of statistical means or sums of situation-specific data that cover various types of situations. The second and the most common type of aggregation takes place when raters who have observed the ratees across situations express their generalizations directly in a nonsituation-specific datum. This type of aggregation is concealed in the answers to traditional inventory questions ("Do you often . . ." etc.), in which respondents aggregate observations across situations, and across situations that they choose themselves. This is also the case, for example, in nonsituation-specific ratings of social behavior, in which situation-specific observations are aggregated by an observer to arrive directly at one aggregate measure for the specific behavior across situations, the relevance of which the observer often chooses. In order to ensure high reliability in observational data, ratings are often aggregated across raters. Aggregate ratings from several observers are thus aggregated twice; first the individual observer has aggregated observations across situations, the properties of which are seldom known and controlled, and second, such aggregated ratings from several observers have been aggregated.

The discussion about aggregation thus far has been concerned with the aggregation of data across situations and data which refer to one and the same, well defined aspect of individual functioning. In this case the operational definition of the psychological phenomena is clear, and the characteristic property of nonsituation-specific data is that they rank order individuals for the trait under consideration without considering differential situational reactions or actions. A special case arises when data are aggregated at the same time across specific behaviors to arrive at aggregate data that are supposed to cover a broader aspect of individual functioning, for example, anxiety, intelligence, dependency. This kind of aggregation is not too uncommon in personality and developmental research. In order to investigate the problem of cross-situational consistency in behavior, data are sometimes aggregated in various ways across specific variables, situations, and time (cf. Mischel and Peake, 1982). This use of aggregation of observations is similar to a meteorologist aggregating data for temperature (in the air, water, and ground), humidity, and wind velocity, all measured across days in order to investigate the consistency of the weather.

Two common hazards in personality and developmental research are connected with the use of data aggregated across specific variables. The first is the tendency toward reification of hypothetical constructs. The reification often leads to implicit assumptions that something, such as anxiety, dependency, intelligence, etc., exists as an independent entity that can be unambiguously defined and measured. One implication of this assumption, which most researchers reject when it is made explicit, is that the factor measured remains the same regardless of how it is measured. Second, and this is a connected point, there is often lacking a conceptual or theoretical basis from which to choose the relevant variables to cover the broader construct. Some researchers may choose certain variables while other researchers may come up with quite another set of variables for the same concept. Thus, though the two researchers intend to elucidate a certain problem concerning the same construct, their empirical results may be contradictory, and yet both are valid.

Of course, nonsituation-specific data can be used for aggregation, depending on the problem under consideration, and they can be used as a basis for appropriate kinds of statistical and multidimensional analyses. However, the basic quality of aggregate data of this type must not be forgotten when interpreting the empirical results. The results of the neglect of the difference in character of situation-specific and nonsituation-specific data can be seen in the use of factor analysis in personality research and the interpretation of associated results. As discussed in another connection, factor analysis of situation-specific data (such as data from intelligence tests) yields factors of a different character from those obtained by factor analysis of non-situation-specific data (Magnusson, 1984a,b). In the first case, using situation-specific data, the analysis homogenizes items into factors on the basis of similarities in the situational conditions (e.g., test items), and in the second case, using nonsituation-specific data, factor analysis homogenizes items into factors in terms of actions and reactions.

METHODOLOGICAL AND RESEARCH STRATEGICAL CONSIDERATIONS IN THE MEASUREMENT OF CHANGE

A crucial and extremely difficult problem in developmental research is the measurement of change (see Harris, 1963; Nesselroade & Baltes, 1979). The central empirical approach to the study of development has been in the study of individual differences. With reference to this fact, the discussion in the following sections is concerned primarily with problems in measuring change in terms of inter-individual and intra-individual differences. Few areas of research are as rife with methodological traps, some of which are difficult to identify and even more difficult to handle in a proper and effective way. This issue has been given special consideration in our longitudinal project (see Bergman, 1971, 1972a,b; Bergman & Magnusson, 1983, 1984a,b; Magnusson & Bergman, 1984a).

The starting point of a discussion on the measurement of change is the existence of individual differences in growth rate for single aspects of individual functioning and patterns of factors. The problem has far-reaching and serious consequences for planning and implementation of developmental research, with respect to cross-sectional vs. longitudinal approaches, and with respect to the choice of appropriate methods for the treatment of data. The methodological consequences of the existence of intra- and inter-individual differences in growth rate have not been given adequate attention in the discussion regarding appropriate approaches to developmental research. It seems worthwhile, therefore, to discuss this topic at some length.

Individual Differences in Growth Rate for Single Aspects of Functioning

In traditional developmental research, the important issue of developmental consistency has more often been confined to the investigations of only one or a few aspects of individual functioning over age stages. Most of the interest has been on the same aspects of behavior across ages, that is, what Kagan (1971) referred to as *homotypic* continuity. Fewer studies have been made of *heterotypic* continuity, i.e., the view that a particular underlying disposition may be manifested in different types of behaviors at different developmental stages, implying that a certain behavior at an early age may be systematically related to other types of behavior at a later age (Moss & Susman, 1980). Bell, Weller and Waldrop (1971) defined these two types of continuity in behavioral terms as "isomorphic" and "paramorphic" relations, respectively.

A common feature of empirical research on homotypic and heterotypic continuity in development, connected with the variable orientation, is the use of linear regression models and methods, particularly correlation coefficients. This is the case not only for the study of homotypic continuity but also for research on heterotypic continuity, as illustrated in the study by Kagan and Moss (1962) on the stability of person characteristics in development from infancy to adulthood.

Linear correlation coefficients are used primarily for four main interrelated purposes in developmental research.

1. For the study of stability of person factors using R-technique (Backteman & Magnusson, 1981; Bayley, 1949; Block, 1971; Bloom, 1964; Olweus, 1979;

Pulkkinen, 1982). Developmental consistency in a certain aspect of individual functioning is assumed to be expressed in terms of the stability of rank orders of individuals for a certain behavior across age stages.

2. In studies of the relations across time among variables in multivariate sets of data when using, for example path analysis or LISREL (Jöreskog & Sörbom, 1981). LISREL enables a researcher to test different theoretical models of development for a certain set of aspects of individual functioning, provided that certain assumptions are met.

3. For construct validation of personality traits. The basic assumption is that high coefficients for the stability of a certain aspect of human functioning across age levels is a prerequisite for claiming that it can be regarded as a trait, for construct validity. Mischel (1968), for example, cited low coefficients for the correlation between measures taken at different ages as an argument against the existence of personality traits.

4. As the basis for the factor-analytic approach to the study of individual development (Lerner, Palermo, Spiro, & Nesselroade, 1982).

An Empirical Illustration

As a background to a discussion, some basic results from an empirical study in IDA can illustrate the application of models of relationships between pairs of variables. (The study is presented in detail in Chapter 6.) The study concentrated on the relationship between individual differences in physical maturation among girls and their later social adjustment.

The relationship between age of menarche for girls and alcohol consumption, when they were 14 years and 5 months as an average (14:5), was first investigated. There was a strong negative relation, reflected in a chi-square that was significant at the promille level. Thus, early maturing girls use significantly more alcohol than late maturing girls at the age of 14:5. At 15:10, the relation of alcohol consumption to age at menarche had decreased to a nonsignificant tendency ($p < 0.20$), while at the age of 26 there was no systematic relation between alcohol consumption and age of menarche. Late and early maturing girls reported almost the same amount and frequency of alcohol consumption at adult age. The results from other studies in the project on social adjustment and social relations during puberty among girls, particularly those connected with norms, values and social relationships, show very clearly that these aspects are also related to individual differences in physical maturity. For some behaviors, the individual differences in social adjustment are temporary and will disappear after a time, as in the case just described. For other factors, individual differences on entering the adult world may lead to more permanent differences in adjustment in adulthood. This pattern is demonstrated in Chapter 6.

These results led to the following conclusions:

1. In all matrices of data for aspects of individual functioning that relate systematically to physical maturity, a portion of the variance will be determined by individual differences with respect to the onset of the physical transitional period. Thus, the coefficients reflecting relationships between these other variables (for example, between aggression and parental relations or between drug abuse and peer relations) will be partly determined by individual differences in physical maturity. This, of course, will be the case even if data for physical maturity are not included in the matrix of data.

2. The strength of the correlation between physical maturity and behavioral variables will vary with the chronological age at which these other variables are measured. This implies that the extent to which individual differences in physical maturity influence the coefficients for relationships between secondary factors in cross-sectional studies will in part depend on chronological age.

These conclusions imply that considering biological age when studying individual differences in social adjustment during puberty might be as important as considering mental age when investigating factors associated with intelligence (see, for example, Magnusson, 1985b; Petersen & Taylor, 1980). Schaie (1965) presented a general model for development, holding that "a response is a function of the age of the organism, the cohort to which the organism belongs, and the time at which measurement occurs" (p. 93). As Baltes (1968) noted, Schaie's model implies that two dimensions are free to vary and that data can be explained by a two-dimensional plane in the three-dimensional space. The present study suggests that adding a fourth dimension, biological age, to the data space will improve explanations of variance in developmental data. It may sometimes be more important to take into account biological age than chronological age.

Individual Differences in Growth Curves

The preceeding findings give rise to the general issue of individual differences in growth curves and their implications for developmental research. Though many researchers have presented and discussed models for growth curves (Baltes, Reese & Lipsitt, 1980; Bateson, 1978; Clarke & Clarke, 1984; Jessor & Jessor, 1977; Loevinger, 1966; Schaie, 1972), the methodological consequences in empirical developmental research of the existence of such individual differences have not received the attention deserved. The problem is illustrated with a few examples related to biological factors for which the growth curves can be confirmed empirically (Magnusson, 1985b).

The first example is the growth curve for the thymus gland. This curve is characteristic for the lymphoid system, including lymph nodes and intestinal lymphoid masses (Tanner, 1978). The typical growth curve for the weight of the thymus gland is shown in Figure 3.1 for four individuals, A, B, C, and D, who differ with respect to biological age. (The weight of the thymus reaches a peak at

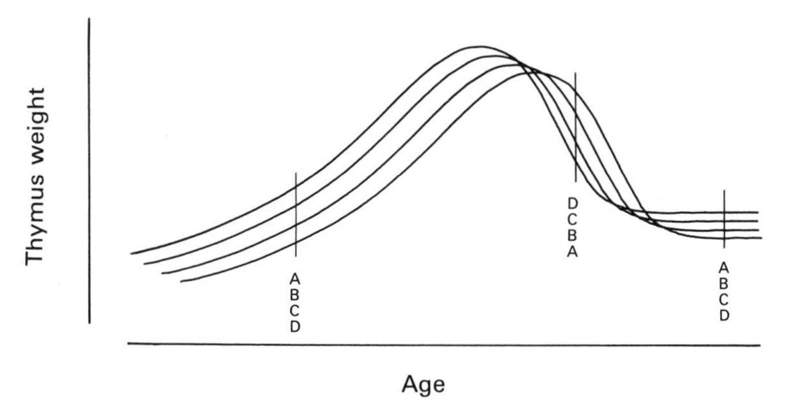

FIGURE 3.1. Fictitious growth curves for four individuals A–D with respect to thymus weight.

around 12 years of age and then declines rather rapidly, to about half that weight at the age of 20.) For various chronological ages, the figure shows the rank order of the individuals with respect to weight of the thymus.

The second example is the growth curve for body height. A large number of studies show that "early-maturing girls tend to be taller earlier in life but ultimately shorter, consistent with an earlier cessation of growth" (Garn, 1980, p. 134) (see Westin-Lindgren, 1979, for an empirical illustration). This model implies that the final level of this person factor is inversely related to chronological age at the onset of the growth spurt. Figure 3.2 shows the growth curves for four girls, who started at the same height but differed in age at the onset of the growth spurt, in concordance with that model. The rank order of girls with respect to height is given for various stages of chronological age.

The growth curves just discussed have been drawn in order to highlight possible methodological implications of the existence of individual differences in physical maturation. Those familiar with psychometrics have already observed that statistical effects are seldom, if ever, as dramatic as those illustrated by the fictitious curves. The extent to which these effects occur depends, of course, on the total individual variation for the function in question. Nevertheless, the effects must be taken seriously and analyzed carefully in order to ascertain the validity of traditional linear psychometric models for meaningful estimations of stability and interrelationships between person functions.

The illustrations presented serve as a background for the following conclusions:

1. Though all individuals pass through the transitional period in the same lawful and predictable way, coefficients for correlations between rank orders of

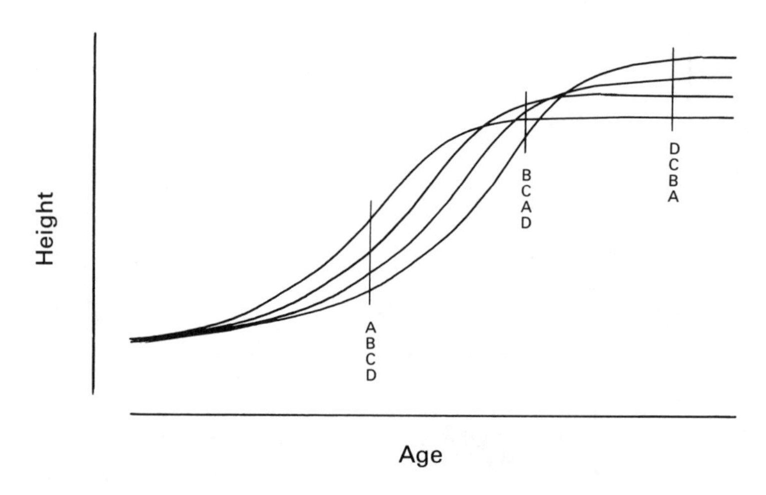

FIGURE 3.2. Fictitious growth curves for four females A–D with respect to height.

individuals obtained for the same function at various age levels can theoretically take both positive and negative values.

2. The size of a coefficient for the correlation between rank orders taken at various age levels depends on the interval between data collections and/or age at the first data collection and the stage in the transitional period at which the data are collected. Coefficients covering the time before and after a transitional period may be higher than coefficients covering a shorter period when data collections occur during the transitional period.

The substantial effects of individual differences in biological maturation can be illustrated with an empirical example from Bloom's review of our longitudinal studies on body height (Bloom, 1964). The coefficients for the correlation between body height at various stages of development and adult body height obtained in these studies were summarized in Figure 3.3a–b.

The coefficients between early and adult body height are lower for the age span from 14–16 years to adulthood than for the longer age span from 8–10 years to adulthood. This can be explained in terms of individual differences in growth rate as discussed above. Figure 3.3b shows that the lowest coefficients are obtained about two years later for boys than for girls, which is in accordance with what is known about gender differences with respect to the onset of the growth spurt.

A second example can be found in a longitudinal study carried out in Finland by Pulkkinen (1982), who observed that prediction of maladjustment factors at

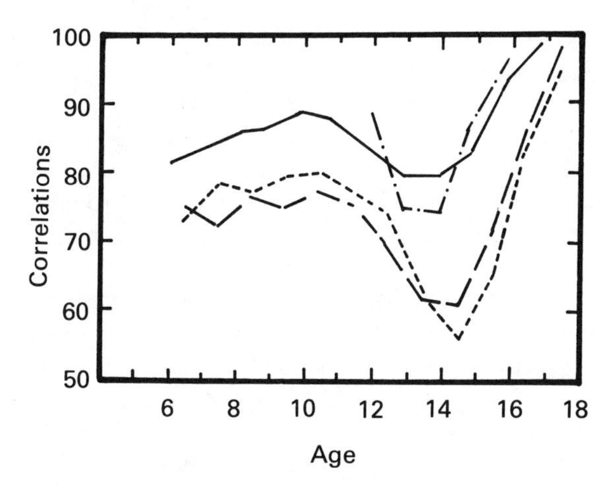

FIGURE 3.3a. Correlations between height at each age and height at maturity of males in four longitudinal studies (adapted from Bloom, 1964).

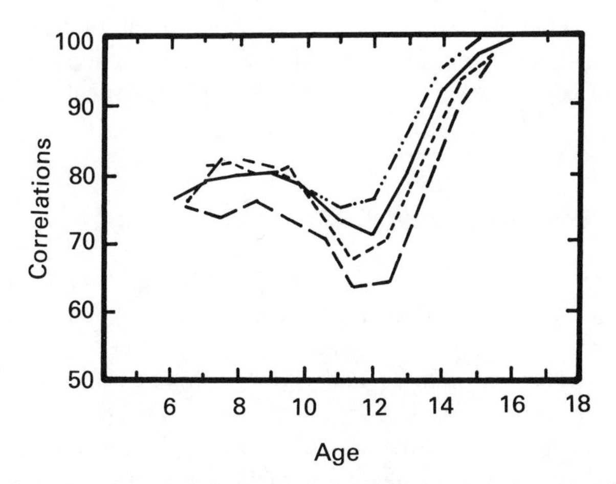

FIGURE 3.3b. Correlations between height at each age and height at maturity of females in four longitudinal studies (adapted from Bloom, 1964).

the age of 20 was stronger for person factors measured at the age of 8 than for person factors measured at the age of 14.

It is worth noting here that what was presented in theory and is empirically demonstrated in Figure 3.3 contradicts and invalidates what has been regarded for a long time as a law of development, summarized here by Clarke and Clarke

(1984): "The second law (as noted) is that, regardless of age, the longer the period over which assessments take place, the lower the correlation is likely to be, that is the greater the change in ordinal position of individuals within a group" (p. 197).

The growth curves in Figures 3.1 and 3.2 represent two examples of biological growth curves. Individual differences in growth rate exist also for many other biological factors, some of them with strong effects on mental functioning and conduct, for example, those involved in the biological changes during puberty. As emphasized, for instance, by Loevinger (1966) in her discussion of four models for individual growth curves, there are many examples of interesting psychological functions that display the same general growth curves as those shown for biological factors (see also Bateson, 1978). In a large scale study Ljung (1965) found evidence for girls and indications for boys of the existence of a mental growth spurt. He also suggested that the mental spurt differs in strength for various factors. These circumstances have important implications for the study of interindividual differences in a developmental perspective.

According to the interactional approach summarized in the preceding chapter, the conclusions drawn for the biological growth curves can be extended to psychological functions. Of particular interest here is the interrelationship between biological and psychological factors. For example the thyroid plays an important role in an individual's general level of activity. Individual differences in growth rate of the thymus gland may then also lead to individual differences in important activity factors and related aspects of psychological development.

Longitudinal Research - Successive Cohorts

A major strategic implication for research of the preceding discussion is that the same individuals must be followed across time if the development of individuals as totalities in a process of maturation and experience in continuous interaction with their environment is to be understood. In other words, longitudinal research must be conducted. The empirical study outlined earlier and the theoretical discussion clearly demonstrate the necessity of longitudinal research for effective analyses of important developmental problems. A longitudinal approach, of course, does not constitute an all-purpose tool in developmental research. But given the nature of the developmental process, seen in an interactional perspective, there is no adequate alternative to longitudinal research for the study of many important developmental issues. To quote McCall (1977): "The longitudinal method is the life blood of developmental psychology: it deserves a more thorough, objective and constructive evaluation by all developmentalists" (p. 341) (see also, Husén, 1981; Livson & Peskin, 1980; Wohlwill, 1970).

Because longitudinal research has become "a la mode", there is a tendency to categorize even rather short-term observations as such. The preceding discussion clearly demonstrates that effective longitudinal studies must cover the total critical periods of development for the function(s) under consideration, and observations must be made continuously during the critical stage of development so that important transitions are not missed. This means in many cases that longitudinal studies must continue over a considerable period in order to avoid misinterpretation of results.

As discussed in Chapter 4, longitudinal research is risky and tedious. One suggested solution has been to study development by using successive cohorts of different ages and starting the observations at the same time for samples of subjects of different ages. Data from the cohorts are then combined in order to investigate stability and change in human development over a wider time span. Even if the role of person-environment interactions is disregarded when comparing data from various cohorts, the discussions above clearly show the limitations of this use of successive cohorts in order to overcome the problems of long-term longitudinal research. The problem connected with interpretation of results from such an approach becomes obvious when the developmental curves for the growth of height in Figure 3.2 are examined. Assume that one cohort has been studied from t_1 to t_2, and another cohort has been studied from t_2 to t_3 in order to investigate the stability of individuals with respect to growth rate in height. Given that, in an extreme case, $r_{t_1 t_2} = 0$ and that $r_{t_2 t_3} = 0$, the most reasonable assumption about $r_{t_1 t_3}$ is that it will also be 0. The correlation in this extreme case would actually be $r = -1.0$. Other cases illustrating the problem with successive cohorts can be constructed from Figure 3.1.

The study presented earlier of the relationship between age at the menarche on the one hand and alcohol consumption at the age of 14 and adulthood, on the other, illustrates empirically how the use of successive cohorts can lead to false conclusions. It is then interesting to note that the empirical coefficients were significant (a) for the relationship between the age at the menarche and alcohol consumption at the age of 14, and (b) for the correlation between alcohol consumption at the age of 14 and alcohol consumption at adulthood. If two successive cohorts had been used, one going from 10 to 14, and the other from 14 to 26, the natural conclusion from these two coefficients would have been that there is a significant relation between age at the menarche and alcohol consumption at adulthood. This is contradictory to the conclusion that can be drawn from the appropriate longitudinal data which showed that this relation was low and insignificant.

The research strategy using successive cohorts in longitudinal research is applicable and useful under certain conditions, but the preceeding theoretical discussion and the empirical illustration show that it must be handled with great care and consideration.

Implications for Treatment of Data in a Variable Approach

The preceding discussion, about the methodological implications of the existence of individual differences in growth rate for single aspects of individual functioning, started by distinguishing four common uses of linear correlation coefficients in developmental research: for the study of the stability of single factors across ages, for the study of relations across time for a multivariate set of variables, for the study of the validity of personality constructs, and as a basis for factor-analytic research.

It should be clear from the foregoing discussion that low coefficients of correlation for the stability of rank orders of individuals across time cannot generally be interpreted as indicators of inconsistency in the person factor under consideration (Baltes & Nesselroade, 1973; Wohlwill, 1973). Nor can they be unreservedly used as arguments against the existence of personality traits or dispositions. Since factor analysis is based on matrices of correlation coefficients, the effects briefly demonstrated here will also occur in factor analytic research on development when R-technique is used, which is frequently the case. These consequences are complex and need careful analysis, particularly considering the complicated process of interaction of biological and psychological factors in the individual and the fact that these interactions may differ in relation to maturational stage, a problem that has been discussed by several developmentalists (McCall, Appelbaum & Hagerty, 1973; Wohlwill, 1980).

Of course, linear correlation coefficients and linear regression models in general are useful tools for developmental research, as they are for other fields of psychology, when properly applied to data in a way that is compatible with the underlying theoretical model. In addition to the traditional linear regression methods (such as path analysis), LISREL methodology can be useful when the necessary assumptions for its use are met. An example of the use of a LISREL methodology is as follows. A longitudinal linear structure model is fitted to a multivariate set of intelligence data, in which the model forms a complete picture of the aspects under consideration. This means, not only that all relevant intelligence components should be included but that there are also reasons to assume that the totality of all the aspects under study can be adequately reflected by the variance-covariance matrix for the variables. This matrix (or the correlation matrix) is the data input in this kind of analysis, and the information used is thus restricted to pair-wise relations. This may be justified by theoretical reasons or by failure to empirically detect higher-order interactions. Certain nonlinear effects and pair-wise interaction effects are, of course, possible to handle using this type of methodology (Kenny & Judd, 1984). There also exist variable oriented methods for studying the multivariate structure of nominally scaled variables such as log linear models (Bishop, Feinberg & Holland, 1975). Within this methodology it is possible to specify models that relate various combinations of the independent variables to observed cell frequencies and then test the fit of different models to data. One problem with this kind of methodology is, however, that even a moderate number of discrete variables that take on a few values may result in empty cells and cause technical complications.

Developmental Analysis in Terms of Patterns

In the preceding analysis, the emphasis was directed to single aspects of individual functioning and relations among them. The discussion in Chapter 1 concerning an interactional model of individual development and the discussion in this chapter on the methodological implications of the existence of individual differences in growth rate emphasize the need for complementary models and methods for analyses that are directed to the study of individuals as totalities.

According to an interactional person approach, the individual as a totality is characterized by the partially specific configuration of relevant, operating psychological and biological factors. The important and interesting changes that take place in the coherent and lawful developmental process are best characterized by changes in such patterns. Thus, models and methods which use individuals as the main conceptual unit for analysis and investigate them in terms of patterns are needed as complements to the regression methods for data treatment which are variable-directed and valuable for analyses of the relations between single variables (cf. Nesselroade & Ford, 1985, and Mumford & Owens, 1984.) A research program using a pattern approach to the study of psychophysiological individuality has been presented by Fahrenberg (1984).

The use of most linear regression models in the study of the relationships among single variables presupposes two assumptions: (a) All variables are applicable to all individuals (Allport, 1937, 1962), and (b) all weights assigned to variables are applicable to all individuals in the sample. These assumptions may well be met in some applications, but their generality can be seriously questioned. It seems more reasonable to assume that there are distinctive, differentially compensatory and stimulating mechanisms, which contribute to the outcomes in specific configurations of factors (cf. Rutter, 1983). This statement refers to the functioning of mental and biological aspects of individual functioning, both in a current and in a developmental perspective, and has important consequences for the study of intra- and interindividual differences when either a cross-sectional and/or a longitudinal approach is applied.

As emphasized earlier the longitudinal program has devoted much time and resources to the problem of studying individuals in terms of their partly specific configurations of data. In general, three types of person-directed methods have been applied (Bergman & Magnusson, 1987).

1. The first set of methods is a cross-sectional classification approach and the procedure is as follows:

a) The individuals are first classified into groups or clusters at each point in time on the basis of their profile of values in the factors relevant at that point in time. Different methods are available under the general heading of *cluster analysis* (Everitt, 1977), which are descriptive. Sophisticated methods based on theoretical models such as latent structure analysis (Lazarsfeld & Henry, 1968) and INDCLUS (Carroll & Arabie, 1983) are also applicable. For applications within the study of development using a person approach a new cluster analytic scheme, RESCLUS, has been developed. A characteristic feature of RESCLUS is that a residue of unclassified individuals are formed (Bergman & Magnusson, 1984a).

b) The pairs of classifications at adjacent points in time (e.g., at the age of 10 and the age 13) are first cross-tabulated (Bergman, 1985; Lienert & Bergman, in press). The significance of the major streams or groupings found can then be tested using ordinary contingency table analysis. It is also possible to form higher order cross-classifications between, for instance, three adjacent points in time.

In this kind of analysis it is important to evaluate and estimate the accuracy of the description obtained, for instance, in terms of percent variance explained. The often noticed sensitivity of cluster analysis to changes in sample composition and method specification is also a matter worthy of consideration. With regard to the first aspect, the above mentioned RESCLUS methodology may be more stable than most methods.

In Bergman and Magnusson (1987), an example is given of the cross-sectional analysis for the study of the development of adjustment problems. An illustration is also given in the study presented here in Chapter 8.

2. The second approach is a longitudinal classification analysis in which the complete set of longitudinal multivariate observations is analyzed in one and the same analysis. One such method, presently being developed in this project, uses longitudinal similarity indexes between each pair of individuals as the input data in a classification analysis.

3. The third set of methods is based on a typological frame of reference, and the aim is to find developmental *types* (i.e., longitudinal value configurations that are in some sense common or more frequent than expected). Some of these methods originate within configural frequency analysis (CFA) (Krauth & Lienert, 1973, 1982; Lienert, 1969; Lienert & zur Oeveste, 1985). The main idea in CFA is to start from the complete multivariate frequency table and look for observed frequencies that are significantly higher than expected under an independence model (types) and observed frequencies that are significantly lower than expected (antitypes). This line of thinking has been extended to more complicated situations including the analysis of longitudinal data. In longisectional interaction structure analysis (LISA), CFA is first applied cross-sectionally and results are then linked over time (Lienert & Bergman, in press). In other

types of methods within this framework, longitudinal patterns are analyzed directly (Kohnen & Lienert, 1987).

PROSPECTIVE AND RETROSPECTIVE APPROACHES TO DEVELOPMENTAL RESEARCH

In Chapter 2 development was discussed in terms of prosimilarity and retrosimilarity of patterns. This brings into focus the distinction between a retrospective and prospective approach to developmental research. When these two approaches are discussed and evaluated, the general conclusion is that a prospective approach is more effective than a retrospective approach, with reference to the deficiencies in retrospective data which have been elucidated in empirical research (cf. Cherry & Rodgers, 1979; Wadsworth, 1979; Yarrow, Campbell & Burton, 1970). However, this negative attitude toward the retrospective approach per se is often due to a confounding of approaches with the type of data used. The efficiency of a prospective versus a retrospective approach is not only in the properties of the approaches in themselves, but also in the type of data that are being used.

Types of Data in Retrospective and Prospective Analyses

The usefulness of data in longitudinal analyses should be evaluated with respect to the following four characteristics (cf. Janson, 1981).

1. *The reference point in time.* The particular occasion or age level in a person's life that a datum refers to.

2. *The coding point in time.* The point in time when the observation is coded into a datum. For example, the observations that a teacher makes about a pupil can be expressed in a datum directly in connection with the observations, or they can be coded 10 years later.

3. *The collection point in time.* The point in time when the datum is collected by the researcher. For example, a register datum for a certain individual may refer to a specific age level and have been coded at the same time that it refers to, but it is collected by the researcher much later.

4. *The usage point in time.* The point in time when the researcher uses the data for calculations.

If the reference point in time and the coding point in time coincide, the datum is not retrospective and may be called prospective, independent of the points in time for collection and usage. If they do not coincide, the datum is retrospective.

Prospective vs. Retrospective Analyses

The distinction between a prospective and a retrospective approach to the analyses of the developmental background to adult functioning and the role of prospective and retrospective data in these analyses will be discussed with reference to an example, in which the factors involved in the developmental process behind adult criminality are the main object of interest.

A prospective approach answers the question: How do individuals, who differ at an early age with respect to factors that can be assumed to be operating in the developmental process, differ at adult age with respect to criminal behavior? The prospective approach takes its starting point in individual differences at an early age in, say, intelligence, social background, and aggressiveness, factors that may be relevant for understanding the developmental process behind adult criminality. For the description of individuals as high or low in these respects, both prospective and retrospective data (as described in the preceding section) can be used. The validity of the assumptions is tested by investigation of the differences in criminal activity in adult age among individuals with different values at an early age in intelligence, social background, and aggressiveness. The analyses can be made for single variables as well as for patterns of variables.

In a retrospective approach to the above problem, the question is: How do individuals who differ with respect to criminal activity at an adult age differ at an early age with respect to operating factors in the developmental process? Thus, the starting point is individual differences in adult criminality. The relevance and importance of single factors or of patterns of factors in the developmental process behind adult criminality can then be investigated by the study of differences with respect to potentially relevant factors at an early age between individuals differing at adult age with respect to criminality. Also in this case analyses can be made in terms of both prospective and retrospective data in principle.

Thus, a prospective and a retrospective approach to developmental problems differ with respect to the questions they answer and the information they yield. Both are useful and effective: the appropriate approach should be chosen with respect to the problem under consideration and the most efficient type of data used (most often prospective data).

With the preceding points in mind, data in IDA are used both for prospective and for retrospective analyses of the problems under consideration. Almost without exception, it is then possible to base the calculations on prospective data.

METHODS OF OBSERVATION

When the Nobel laureate Wigner (1969) stated that modern microphysics and macrophysics no longer deal with "relations among observables but only with relationships among observations," he hit on one of the core issues of empirical psychology as well. The declaration emphasizes a point of crucial importance for the interpretation of empirical results in psychological research, namely, the relationship among phenomena, concepts, observations, and data. This issue has been discussed by researchers for a long time, but nevertheless it has been neglected too often resulting in much confusion and misunderstanding in the interpretation of empirical research.

The interpretation of the results of empirical studies of psychological phenomena is usually based on data. The extent to which data really reflect the phenomena—structures and processes—that are under investigation is decisive for the validity and usefulness of the interpretation. For the study of some aspects of behavior, for example certain molecular physiological reactions, this does not cause any serious problems. The measurements directly reflect the concrete feature under consideration without too high a risk of distortion, and the results can be interpreted with minimal uncertainty. For some variables the situation is different; they are the most frequent in psychological research.

The implications of the reasoning above are particularly important when we are dealing with theoretical concepts that are inferential in nature, for example, distraction, motivation, independence, helplessness, intelligence, needs, and motives. In these cases, whatever the theoretical conceptualization of these variables may be, what we are actually studying and can interpret are relationships among observations. The crucial issue then becomes how well the relationships observed in the data reflect the relationships in the space of phenomena, and this fit between data and phenomena depends on our methods for observation. This makes choosing the appropriate methods for observation a central issue in the planning of psychological research.

The preceding points underscore the necessity of paying attention to the methods for observation in any kind of psychological research. In planning the various stages of IDA much time has been devoted to this issue. Most of the methods for observation and the psychometric properties of the data obtained have been described in the preceding volume on this project, and for details the reader is referred to Magnusson, Dunér, and Zetterblom (1975) and Magnusson and Dunér (1981).

The dominant mechanistic and reductionistic view of psychological research has exhibited a strong tendency toward *methodological monism* (Rychlak, 1981; Toulmin, 1981). In the history of psychology, the experiment has been preferred as the "scientific" method. Even though other methods were viewed as theoretically possible and hesitantly allowed, the experiment was regarded as the true scientific tool for research. This circumstance might be one reason that some aspects of individual functioning (e.g., individuals' basic value systems) that are essential for the understanding of the functioning of the totality, but cannot easily be made the object of experimental research, were omitted from the phenomena being studied. The bias is reflected in the fact that the main field of psychology was named "experimental psychology" after a methodological approach, not after its main content. However, it is well known that there is nothing that is, once and for all, *the* scientific method. The truly scientific attitude is that the relevance of a scientific method for empirical research on a certain phenomena can only be judged with reference to the nature of the problem under empirical study. As Toulmin (1981) claims, it is paramount to exchange *methodological pluralism* for methodological monism. The only criterion for what is a scientific method for observation is the extent to which it helps to solve our problems in an effective way. In 1968, Peterson drew the general conclusion that the best way to obtain information pertinent to change is to employ the full range of methods in behavioral science. For the planning of IDA the implication of this view has meant that all possible methods of observation and data collection have been used with reference to the specific character of the problem under consideration.

Chapter 4
PLANNING AND IMPLEMENTATION OF THE PROJECT

INTRODUCTION

The view presented in the preceding chapters led to a series of implications for the planning of IDA. At various levels of generalization these implications follow implicitly from the theoretical formulations. In the introduction to this chapter some general implications of special importance for the planning, implementation, and interpretation of the longitudinal project are summarized and discussed as a background to the presentation of the project with respect to research strategies and administration of data collection. Generally, these implications were explicitly formulated from the beginning and provided the basic framework for the general design of the project and the first stages of data collection and treatment (Magnusson & Beckne, 1967; Magnusson & Dunér, 1967; Magnusson, Dunér, & Beckne, 1965; Magnusson, Dunér, & Olofsson, 1968; Magnusson, Dunér, & Zetterblom, 1968). This theoretical framework has been refined as the project has evolved, drawing from the international debate on developmental theories, increased knowledge offered by empirical developmental research, and the experience gathered in our own basic research on person-situation interactions as well as from the longitudinal project. This evolution is a natural product of the scientific process.

Research Strategy and Methodological Implications of the Theoretical Framework for IDA

A *Longitudinal Approach.* A primary and essential consequence of the general view presented in the preceding chapters was the decision to study the developmental background of adult individual functioning by following the same individuals across time, that is, by longitudinal research. It was also clear that the same individuals had to be studied from childhood to adulthood in order to avoid the limitations of successive cohorts. Adoption of a longitudinal approach imposes heavy costs in time, money, and commitment, but there is no way to avoid these costs, if we are ever to gain the knowledge we seek.

The longitudinal approach was influenced by the view of "normal" individual development as an adaptation process in which the individual is a conscious, active agent. The planning of the study of normal development formed the framework for the study of the developmental background of adult maladjustment. For the fullfillment of that purpose a longitudinal approach has decisive implications. In his discussion of research on the development of crime and delinquency, Farrington (1979) summarized the importance of conducting longitudinal research for the elucidation of the following issues:

1. Prevalence of crimes and convictions
2. Peak age for arrests and convictions
3. Delinquent generations
4. Juvenile delinquency and adult crime
5. Transition probabilities in criminal careers
6. Specialization in criminal careers
7. Prediction of the onset of convictions
8. Prediction of recidivism
9. Effects of penal treatment
10. Effects of events on development
11. Testing biological theories of etiology
12. Transmissions of criminality between generations

Some of these issues are restricted to research on the development of crime and delinquency. However, some are also relevant for the planning of research on other aspects of maladjustment, such as alcohol and drug problems, psychological disorders, and mental illness.

The planning of the study of individual development as a longitudinal approach did not, of course, imply a negative evaluation of cross-sectional studies per se nor preclude the use of cross-sectional studies of developmental issues. The two approaches cannot be substituted for each other; they are complementary. As illustrated in Part II, which presents empirical studies, the data have been used for both cross-sectional and longitudinal studies in IDA. The complementarity of cross-sectional and longitudinal research is not symmetrical, however. Data collected cross-sectionally within a longitudinal design may be used to answer questions about current relationships. However, data collected within a purely cross-sectional design are often of limited value for the elucidation of developmental processes.

Control of Cohort Effects. Factors at various levels of the environment are involved in the interaction process between an individual and the environment. A cohort of subjects that is homogeneous with respect to age will interact with an environment that is different in various respects from the environment with which other generations have interacted. Such differences in the environment can influence the results for a certain aspect of individual functioning and should be controlled (Baltes, Cornelius, & Nesselroade, 1979). Consequently, the project included more than one cohort to permit the estimation of the size of cohort effects and to permit the control of such effects, when appropriate.

The extent to which cohort effects may influence the results depends, of course, on the nature of the problem to be analyzed, the level of the processes, and the types of factors involved. The main focus of the project is the lawfulness of the processes underlying individual development. Central problems can then be studied effectively without controlling cohort effects. For example, the lawfulness of the processes relating neurotransmitters and conduct, or conduct and social relations, can be effectively investigated without consideration of cohort effects. In other cases, such as the youngster's choice of educational and vocational careers, the results are influenced by the specific conditions in the labor market at the time young people make their choices. The interpretation of the results should then be made with reference to these conditions, and it is considerably enriched by information about cohort effects in relevant respects.

The Importance of Representative Samples. An important prerequisite to both the study of "normal" development and the developmental background of adult maladjustment is that observations are based on a representative group of subjects. Results obtained from groups of subjects that are systematically selected according to specific criteria and that are known to deviate from the "average" can be useful and valuable because they can be interpreted with reference to the specific character of the selection variables. However, many of the results published in developmental research have been obtained without consideration for the representativeness of the groups studied. This is particularly true for research concerned with various aspects of maladjustment. Striking illustrations of this problem can, for example, be found in the reports published on the hyperactivity syndrome. Few studies in this field have been presented in which representative groups of sufficient size have been followed across time according to a prospective plan for the study of hyperactivity in a developmental perspective.

In IDA, the cohorts studied were composed of representative samples of boys and girls. The way the cohorts were defined and actually composed contributed to representativeness in most relevant aspects.

The Importance of Using Methods for Observation Appropriate to the Level of Processes and to the Type of Aspect Being Studied. As emphasized in Chapters 1 through 3, the total process in which an individual functions and develops includes processes that involve systems of operating factors at various levels of complexity. If the analyses of these processes and structures are to be successful and effective, the data used for any given analysis must be appropriate for the level of complexity and the type of specific structure and process under consideration.

The type of data used on a particular occasion is actually decided by the choice of procedures and methods for data collection. A considerable amount of work has been devoted to matching data collection to the problem under consideration. Since the ambition has been to cover a broad spectrum of aspects of individual functioning at the various levels of processes and structures involved, the consequence of the forementioned reasoning was that a wide range of available methods for observation and data collection have been used.

The instruments and procedures used for data collection had to meet definite criteria in order to be appropriate for use with a representative sample of individuals. Among other things, it implies the requirement that instructions and tasks should be interpreted in the same way by all respondents, independent of intelligence level, as far as possible. This requirement raised some problems for the use of inventories. Certain available inventories were identified to be appropriate for the project, such as inventories for personality factors (KSP), interests, Type A behavior, and femininity-masculinity. However, for many of the central interests in the project, including psychosomatic reactions, attitudes towards school, norms and norm breaks, values, further educational and vocational plans, leisure time activities, alcohol and drug abuse, and criminal activities, no satisfactory methods were available and measures had to be constructed with consideration to the requirements discussed here.

Most of the data collected and used were designated as "nonsituation-specific" data, that is, data that reflect differences in individual general disposition to act and react across situations. Nonsituation-specific data are relevant for the study of the role of such characteristics in the reciprocal process of interaction between the individual and the environment. However, as was argued in Chapter 3, such data should be complemented with what was designated as "situation-specific" data, that is, data that reflect individual actions and reactions in specific situations or types of situations. For that purpose data were also collected in specific situations, which makes it possible to relate individuals' actions and reactions to specific known and controlled situational conditions. The importance of including this type of data is illustrated in the study presented in Chapter 7.

In the experimental tradition a favored approach is to break down individual functioning into the smallest unit possible for investigation. In this tradition, ratings covering broad aspects of, say, conduct or social relations, are regarded as less scientific, if tenable at all, for psychological investigations. During the sixties and seventies a strong criticism was launched against the appropriateness of behavior ratings by personality researchers (Berman & Kenny, 1976; D'Andrade, 1974; Mulaik, 1964; Passini & Norman, 1966; Schweder, 1975). The argument raised by them has been discussed and countered by others (Block, 1977; Block, Weiss, & Thorne, 1979; Goldberg, 1978; Backteman & Magnusson, 1981). It has also been counteracted by empirical data from the longitudinal project as illustrated, for example, in the empirical studies presented in Chapters 6 to 8.

With reference to the discussions in Chapter 3, two comments on this issue can be made. First, the structures and processes involved in, for example, conduct and social relations are complex. Many single factors operate together in a complicated, often nonlinear, way. Second, these complex processes cannot always without further consideration, be understood and explained in terms of some linear sum or other function of a number of single factors involved. In order to understand and explain conduct and social relations, for example, it is necessary to use a procedure that provides data congruent with the level of complexity of the phenomena under study. Rating procedures will yield the appropriate data for many important aspects of individual functioning that are of central interest for understanding the total lawful process of development. Scarr (1981) has well formulated the standpoint that formed the basis for the decision to rely heavily on ratings in the planning of the longitudinal project: "Behavior ratings by people who know the child well, who have observed and interacted with him in a variety of settings, will be worth far more than the most extensive counts of discrete behaviors observed for brief periods in one or two settings" (p. 164). Rating procedures were judged to be particularly appropriate for the study of the two areas of interest that were used as examples above; in investigations of conduct and in the study of social relations (cf. Cairns & Green, 1979).

Implications for IDA with Respect to Content of the Theoretical Framework

A Broad Spectrum of Relevant Factors. A central element of the general theoretical background for IDA was the view that an individual develops as a psychological and biological totality in which each aspect derives its functional meaning from its role in the totality, and therefore that development is characterized by individually, partially specific patterns of presumably relevant factors in the individual and the environment. The implication of this view for the planning of the project was that a broad range of factors that were assumed on theoretical grounds to be operating in the developmental process had to be covered. This general view was particularly relevant for the study of the developmental background of adult maladjustment. On the person side this implied the inclusion of five main categories of factors: (a) cognitive-affective factors (self-perceptions, values, goals, strategies, intelligence, creativity, self-evaluations, sense of control of own life, emotions), (b) personality or temperament factors, (c) biological factors, (d) conduct, and (e) social relations. Four of these categories have been extensively investigated in traditional developmental research but as emphasized in Chapter 1, biological factors have not been given the attention they deserve (Petersen & Taylor, 1980). With the general background presented in Chapter 1 about the important role of such factors for the total process of interaction among subsystems in the individual and thereby for the individual's functioning in interaction with the environment, a cornerstone in the planning of IDA was the inclusion of biological factors.

On the environmental side, the adopted view required the inclusion of factors describing each individual's actual physical and social environment; the physical, economic, and educational standard of upbringing conditions, the physical environment for leisure-time activities, the school environment, and the main aspects of the social environment—family relations, peer network, and peer relations. Above all the interest in the environment was directed toward the individual's own perception and evaluation of his/her network in the family and among peers and toward educational and vocational plans.

The circumstance that a broad spectrum of factors was covered concerning the individual and the environment has had important side effects. Besides the study of issues that were formulated at the onset, it enabled the investigation of new problems that had not been foreseen. Some of these were a consequence of findings in the project. Other new issues arose as a result of studies published by others. The advantage of including these studies in the project has been that they could be planned and the results interpreted with reference to results in other relevant aspects for the same individuals.

The Choice of Aspects with Reference to Age. An important implication of the theoretical perspective has to do with the choice of methods for data collection at each level.

Much developmental research, particularly research directed toward the issue of stability and change in development, has been variable-oriented, that is, directed toward single aspects of individual functioning or clusters of aspects, such as intelligence, aggression, attachment, and social relations. The general approach then, has been to use the same kind of instrument for the collection of data at various age levels in order to ensure that what is assumed to be "the same" person factor(s) is/are measured. The question of what constitutes "the same" function is the well-known problem of equivalence of functional measures, a problem that has important conceptual and methodological implications (see Baltes, Reese, & Lipsitt, 1980; Blalock, 1982; Loevinger, 1966). By contrast, for each age level it has been our strategy to maximize information about the main aspects of the total patterning of operating factors characteristic of that specific developmental period. For example, in studying the early adolescent stage, particular interest was devoted to the individuals' social relations; the character of the social network, relations with peers and parents, among other factors, with reference to the essential role of these aspects during that period.

Aspects of individual functioning and methods used for observation were chosen following the analysis of available knowledge and existing developmental theories. Consequently, the methods used to observe and describe the phenomena under consideration at each age level were chosen with reference to the role of various psychological factors for that specific age level, independent of whether they were used at other levels before or after.

Protective, Moderating Factors. A central theme in the project is the developmental process underlying social adjustment. Because most research on environmental factors in the developmental process underlying adult maladjustment as manifested in alcoholism, criminality, and mental illness has been directed to the investigation of so called risk factors, less interest has been devoted to the search for factors that moderate possible negative factors and thereby prohibit negative development in children who grow up in such environments. The planning of the longitudinal project included analyses of relevant aspects of the environment that could serve as protective factors. Two main hypotheses concerning the environment were crucial for the planning.

First, a physical environment that offers active boys and girls opportunities for constructive and meaningful activities within the boundaries of the law during their leisure time may keep them away from destructive unlawful activities (cf. Barker & Associates, 1978). This assumption led to a careful socio-ecological analysis of the physical environment for each individual child with respect to home conditions, opportunity for leisure time activities, the school environment, and other relevant aspects at an early stage of the project.

The second hypothesis was whether or not a boy or a girl participates in drinking, drug abuses, and criminal activities depends, among other things, on the content and strength of their integrated system of norms and values. For the development of norm systems, both peers and adults, particularly the parents, play a central role. In their recent book, Wilson and Herrnstein (1985) discussed two main dimensions of parental practice as particularly important for the socialization process of a child; the degree of warmth, affection, and support that parents provide their children and the firmness and consistency with which they enforce rules of behavior. They argue that delinquents are unlikely to have had restrictive and warm parents but are likely to have had parents who were emotionally distant, did not care for their children, and disciplined with inconsistent, unpredictable punishment. In the planning of IDA this hypothesis meant that particular interest was devoted to the study of relations with peers and adults during the adolescent period, when much of the norm systems and their attached values is stabilizing.

The second hypothesis was tentatively tested in a cross sectional analysis of data from the project (Dunér & Magnusson, 1979). As expected there was a significant, negative relation between positive relations in the family and the risk for negative social adjustment in the boys. The relation was stronger for boys from families with low socioeconomic conditions. This result was revealed in both linear and nonlinear analyses of data. Good relations with parents were coupled with low risk for negative social adjustment for boys, independent of the socioeconomic conditions, while bad relations to parents were strongly related to high risk for negative social adjustment among boys from low socioeconomic conditions but not to the same extent for boys from high socioeconomic conditions. The statistical interaction revealed in these results illustrates well the basic proposition in the general theoretical framework of IDA. Family relations or socioeconomic factors in the family background per se are not the important factors that are effective in the socialization process of the youngsters. It is the combined, joint effect that is important.

The results are in line with those reported by Werner and Smith (1982) (see also Garmezy, 1981; Garmezy & Nuechterlein, 1972; Garmezy & Tellegen, 1984). The findings suggest that positive relations with parents can serve as an important protective factor in environments that otherwise indicate a risk for negative social adjustment.

SUBJECTS

Sampling of Subjects

One primary question in the planning of IDA concerned the appropriate sample(s) in relation to the goal of the project. With reference to the viewpoints presented in the preceding chapters the final decision about the choice of sample(s) to be studied was based on the following requirements.

1. The sample(s) should be representative, as far as possible, of the age group(s) to be covered. For example, the full range of intelligence and social background should be included.
2. The total sample(s) should be large enough to permit generalizations about the age group(s) for the features of individual functioning that were to be studied.
3. The sample should be large enough to form a frame of reference for the study of groups of children who are homogeneous with respect to some characteristic(s) (e.g., isolated and/or rejected children), using the total group as a frame of reference for the interpretation of results for each homogeneous group.

4. The sample should be large enough to permit the separation of the total group into two random subgroups, for the cross validation of results.

5. The sample(s) should be small enough to permit effective procedures for data collection with as little dropout as possible and effective control of dropout.

Two options were possible in order to meet these requirements: (1) to draw a random sample of sufficient size from the total national population of youngsters of each age and, (2) to use all boys and girls in one community as the sample, assuming that it would be representative for youngsters of the same age.

The disadvantages with the first option are severe for carrying out a longitudinal project across any significant time period. First, the students could not be studied as a group without substantial cost and administrative problems as they are spread over the country. Second, to find and study one individual at a time for variables that can only be studied by methods that are administered individually (for example, laboratory measures), would be very costly in time and money. Third, the follow-up of these individuals would require, on each occasion, the necessary investment of great resources.

Given these considerations the second alternative, to study all boys and girls of the relevant ages in one community, was chosen.

The Community

After careful consideration, a community in middle Sweden, with about 100,000 inhabitants, here called Swe-town, was chosen for the study. Swe-town is an old town that has expanded at a normal rate. It was chosen for several reasons. Initial investigations showed that the community could provide a sample of children reasonably representative of the study's population of children, a finding that has been confirmed in studies of intelligence and socio-economic status of the families (Bergman, 1973). The community includes an urban and a rural area. It has a well-developed educational system, (elementary school to university education), and has a diversified industry. An important practical consideration was that the local school authorities, the teacher organizations, and the parent association were positively inclined to the project and offered their cooperation in planning and implementation.

Cohorts

Three Cohorts. The purpose of the project has been fulfilled by the study of three cohorts of subjects. They consisted of all boys and girls, who attended school in Swe-town and belonged to the third, sixth, and eighth grades respectively during the first data collection in 1965. Thus the majority of the subjects were 10, 13, and 15 years of age for the respective cohorts.

TABLE 4.1.
Number of Subjects in the Three Cohorts at the First Data Collection (1965)

Grade	*Approximate age*	*Number*
3	10	1,025
6	13	960
8	15	1,259

The oldest of the age groups was investigated only at age 15. It has mainly served as a pilot group. The second cohort was followed from age 13 to the age of 23, with consecutive data collections during the time at school and one main data collection at adult age. This group is designated the *initial* longitudinal group. The third group, which was 10 years of age at the start, has served as the *main* group for investigation. This group has been followed continuously during the school years and a comprehensive data collection was carried out when they were 26 to 28 years of age. The data collections will be described more fully later in this chapter.

Number of Subjects. The number of pupils in the three cohorts during the first data collections are reported in Table 4.1.

About 89% of the children were born in 1955, 1952, and 1950, respectively; about 10% were born in the year before these "normal" years of birth; about 1% were under-age, having been born in the year after the majority in the grade. As the principles for judging school readiness were not changed between the years in which these groups started school, each of the grades, with the exceptions given below, comprised an age stratum of the whole population of Swe-town.

The following categories were excluded from the data collections at school:

1. Children with serious intellectual shortcomings, that is, with an intellectual level corresponding to an IQ of 70 or lower. Such children are transferred to a special school during their first school year.
2. Children with grave physical handicaps, blindness for example.
3. Children with serious social adjustment disturbances, taken into custody by society and placed in foster homes or school homes outside Swe-town. Students who, after a total assessment, have been placed in such homes have been included in follow-up studies.

Custody of the type mentioned in point 3 is extremely rare with respect to children during their ninth year. Exclusion of this category during the initial investigation did not affect the composition of the main group to any significant extent.

It should be noted that there are almost no private schools in Sweden and none in the area of Swe-town. Children from all social groups attend the compulsory school.

Two points concerning the number of subjects who participated and are included in the calculation of data should be mentioned. First, during the implementation of the project at the time when the subjects attended school, all those boys and girls who entered the school system and belonged to the cohorts under consideration were added to them. Second, in the follow-up at adult age of the main cohort, all those who had participated in at least one data collection during the time at the compulsory school were included. As a consequence of this procedure the main group studied in the follow-up at adult age (those who were ten at the start) consisted of 1,393 subjects, 710 males and 683 females.

Comment. The general research strategy using three cohorts has advantages in three main respects:

1. The administrative procedures and instruments for data collection and so forth, may be tested on the pilot group and the initial group as a basis for possible revision in the planning of data collections for the main group.
2. It permits, to some extent, the control of cohort effects of the kind discussed earlier, when such control is important. In the present case cohort effects can only be controlled for a limited time.
3. The cohort strategy allows the investigation of possible testing effects, by providing panel data for comparison among subjects of the same chronological age but with varying degrees of testing experiences (cf. Berglund, 1962). It should be observed that testing effects may be confounded with cohort effects of other kinds.

Longitudinal research entails the investment of large financial resources, time, and personal commitment. Because the outcome at various stages of the process depends on what has or has not been done at earlier stages, even small mistakes at an early stage can have long-lasting, serious effects. In experimental laboratory research the consequences of a mistake in the planning or implementation of a study may have serious consequences but, normally, mistakes can be corrected by starting again if necessary. In longitudinal research one has to live and deal with the mistakes and omissions that have been made in the preceding process. The use of the pilot and initial cohorts, on which administrative procedures, methods for data collections, and results for important factors, have been tested, has provided invaluable experience, especially about local conditions, which has been of great importance in planning the investigations of the main cohort, both as to methodology and content.

The fact that the cohorts studied represent the full range of person characteristics has three advantages that are essential in fulfilling the purpose of the project.

1. For some aspects of individual functioning, information about the prevalence of various aspects of individual functioning, for example, conduct, criminality, alcohol and drug abuse, and mental problems at various age levels are of central interest (cf. McGee, Silva, & Williams, 1984; Rutter, Cox, Tupling, Berger, & Yule, 1975; Vikan, 1985). Such information is less meaningful if it is not obtained from representative samples.

2. A representative sample implies that the data contain the total variance in the population. This fact makes it possible to use the traditional set of statistical procedures, including regression models, and interpret the results without unreliable corrections for restriction of range.

3. A major problem in developmental research is dropout. Generally, this becomes a problem from the moment the sample is drawn; for various reasons some of the expected subjects do not participate. In many cases, particularly in the study of conduct problems and other aspects of maladjustment, this constitutes a crucial problem, because those first to dropout may be the very individuals of greatest interest. In IDA, the loss of information with respect to data on important psychological and biological characteristics as well as on family background conditions from the early years at school is almost negligable. An effect of this is that it makes it possible to evaluate the potential biasing effect of the attrition that has occurred later, in the longitudinal follow-up of the subjects.

EXTENSIVE VERSUS INTENSIVE STUDIES

With regard to group size and research strategy, two types of investigations have been performed: total group investigations with extensive data collections, and small group studies, in which extensively collected data have been complemented with intensive data collections. Total group investigations have been performed on all three cohorts, and sample investigations have been done only on the main cohort.

Total Group Investigations

Data for the total group have been collected mainly in two ways: (a) With methods that permit administration of tests, inventories, and rating procedures to the total group. Data were obtained in this way from the subjects, teachers and parents. (b) From official records. Data from official records have been obtained for pre- and perinatal factors, family characteristics, alcohol problems, criminal activities, and psychiatric care.

Data from the total group investigations have served two purposes. First, they have been used for analyses of characteristics of the total group. Second, they have formed the frame of reference for the formation of groups of individuals to be studied intensively and for the interpretation of the empirical results from such intensive studies. The results from a sample of subjects may then be interpreted in the frame of reference of data of the total group. This increases the value of the sample results. The studies presented in Chapters 6, 7, and 8 are based on extensive data of the type described above.

Small-Group Intensive Studies

Some of the main problems of individual development cannot be studied effectively only using procedures that can be administered collectively to large groups. For the elucidation of such problems, complementary data are needed that are collected individually by laboratory methods, interviews, certain types of tests, and so forth. Because such methods are expensive and time consuming, data collections using these methods were planned and tested on smaller groups of subjects, taken from the total group.

Two types of intensive studies on groups of subjects taken from the total cohort have been and can be implemented: (a) sample studies and (b) studies of homogeneous groups.

A *Sample Investigation.* As emphasized in the discussion of individual development in the preceding chapters, biological aspects of individual functioning play an important role in the interaction processes. Individual data for such factors were collected from a sample of about 250 boys and girls of the main cohort. When they were between 12 and 13 years of age, the children participated in the following data collections:

1. A test of electrophysiological activity in the brain (EEG).
2. Collection of samples of urine for the investigation of catecholamines (adrenaline and noradrenaline) in two different situations.
3. A test of physical strength and lung capacity measured by ergometer cycling.

On a later occasion, data were collected for *bone ossification*.

In the follow-up of the main group at adult age, this biological sample participated in complementary, intensive studies. These concerned both biological and psychological person factors.

A *Study of a Homogeneous Group.* An illustrative example of intensive studies on small, well-defined homogeneous groups is the study of boys and girls from the main cohort who were neglected and/or rejected by their peers at an early age (cf. Magnusson, Dunér, & Zetterblom, 1975, pp. 225–243).

After a year of successive observations in order to ensure correct identification of neglected and/or rejected children, the 15 girls and 15 boys who best met the criteria, were selected as the experiment groups. Two control groups for each sex were formed in the following way. For each boy and each girl in the experimental groups, the most popular boy and girl, and a boy and a girl chosen at random, both from the same class as the one in the experiment group, were selected. Data collection from the 45 boys and 45 girls who made up the experimental and the control groups, included (a) separate, independent interviews with the two parents in their homes, (b) ratings of the boys and girls based on systematic observations by the teachers in the classroom and in the school yard during a one month period, (c) interviews with the children, performed by school psychologists, and (d) a medical examination. The further course of life of the boys and girls who participated in this study is now being investigated with particular emphasis on the role of early social neglect and/or rejection (Zettergren, 1979; 1980; Zettergren & Dunér, 1979).

DATA COLLECTION

In total, the data collections for the main group have yielded data that may be categorized in the following way:

1. Data obtained from the subjects, parents, and teachers during the time at school.
2. Data obtained directly from the subjects at adult age.
3. Data from official records.

The main characteristics of the collections of data for the three categories are presented in the following. Because the comprehensive procedures for data collection at school were described in detail by Magnusson et al. (1975), the emphasis here is on the second and third types of data collections.

Data Obtained from the Subjects, Parents, and Teachers During the Time at School

The factors that were investigated, procedures used for data collection, instruments used, and the main psychometric properties of the data that were obtained in data collections at school were presented and discussed by Magnusson and Dunér (1981), and by Magnusson et al. (1975). An overview of the variables investigated are presented in Table 4.2.

TABLE 4.2.
Variables Covered by Data for the Main Cohort During the Period at School (10–19 Years of Age).

Conditions of the Home
Type and material standard of the home
Parents' education
Family constellation
Parents' engagement outside the home
Parents' work conditions

Mental capacity
General intellectual ability
Creative ability

Achievement
Grades
Scholastic achievement
School motivation
Relative achievement
Educational and vocational plans and aspirations

Attitudes-interests-evaluations
Self concept—self perceptions
Sex-role identification
Interest domains (social, technical, etc)
Sense of control of own life
Life values
Norms: Evaluations; Behavior intentions; Sanctions; Norm transmitters
Self-confidence

Behavior
Conduct: Aggressiveness; Shyness; Motor restlessness; Lack of concentration; School motivation; Disharmony; Tension (Overambition);
Norm violations
Delinquency
Alcohol and drug use
Social timidity
Leisure time activities
Emotions and temperament
Behavior at home

Biology
General physical capacity
Neurophysiological condition; EEG
Hormonal activity/reactivity: Adrenaline, noradrenaline
Biological age: Ossification; Age at menarche

Social relations
Relations with father
Relations with mother
Relations with peers
Dating

Administration of Data Collections at School. The success of longitudinal research depends on the continuous cooperation of many individuals over many testing occasions. Forming and maintaining cooperative relations with those directly involved, the subjects themselves, the parents, the teachers, and the school authorities, as well as other relevant groups (e.g., the press) are imperative.

For a number of years the data concerned children and were collected in schools and in homes. This method required continuous cooperation with pupils, parents, teachers, and school authorities. Much work was therefore devoted to information and communication with these groups. Measures taken in this context were tailored to the groups and problems involved.

An Advisory Group. A central role in the planning of every stage of the data collections was played by an advisory group that was formed at the beginning of the project and served during the entire period of data collection in the school system of Swe-town. The composition of the group was surprisingly stable, affected only by changes in professional positions of individual members.

The advisory group consisted of one member from the Swedish National Board of Education, one member from the local board of education (the chairman), the head of the local school system, one school principal, the chief medical person, the chief psychologist, one member of the parent association (the chairman), and one teacher from each of the three main grade groupings in the school system.

At an early stage of the planning of each data collection, the preliminary material to be distributed was discussed in meetings with the advisory group. The contact with administrative representatives made it possible to discuss administrative limitations and ways to handle them properly. Teachers and parents provided essential information about attitudes and reactions toward the form and content of the methods used. Members of the advisory group participated very actively and constructively and contributed substantially to the success of the data collections.

Information for the Subjects. The most important persons to be kept informed were, of course, the subjects. Before each data collection at school they were informed about the general purpose of the project. Besides the specific information given in the instructions for the various data collection instruments, the pupils were regularly invited to discuss their reactions to the specific procedures and the project in general.

Information for Parents and Teachers. Besides exerting a direct influence on the procedures and instruments for data collection, the advisory group provided direct contact with school authorities, teachers' organizations, and the parents' organization. Continuous written information was given to parents beforehand in connection with the parts of the data collections that did not concern internal school variables. All teachers were also informed beforehand, in close cooperation with the school authorities. On each data collection occasion, one member of the research group was present at each school, ready to answer questions from teachers and to present supplementary information. The teachers were presented with full information about tests, inventories, and procedures.

Cooperation with the Press. In a large-scale data collection running over a rather long period of time, there are many possibilities for misunderstandings and conflicts. Stories are told and rumors circulate about procedures and the specific or general content of certain instruments. Many examples show how this has caused serious trouble in longitudinal projects.

The press can play an important role in this area. It is the duty of the press to observe and scrutinize what is going on in society and to present its findings to the public. However, it has happened that the press has published information about research projects that has been based on misinformation or misunderstanding of facts, whereupon chances of completing the projects have been severely damaged. Such incidents have often led to a critical and suspicious attitude toward the press among researchers in the behavioral and social sciences, and many researchers have tried to avoid giving the press access to their instruments.

In contradiction to the dominant tradition in this respect, the opposite approach was taken in our project. From the beginning, we took the initiative in cooperating closely with the two local papers. The editors of the papers were given full information about each stage of the project, including all tests and instruments, and each paper assigned a reporter to follow the project in cooperation with the researchers. Material about the project was distributed continuously to the papers for publication prior to phases of the extensive data collections. Short articles by the researchers about results judged to have no biasing effect on further data collections were distributed as well.

The cooperation with the press was very successful. In no way did the papers misuse the information. On the contrary, their regular publicity supported the project. On some occasions the journalists, who had access to full information, could inform parents who had contacted them that the information they had received from their children about the project was not correct.

Comments. The handling of information and communication problems took more time and resources than expected. However, this work was considered to be a necessary prerequisite for good results in large-scale longitudinal research, and was worthwhile. For example, in the first stage of the project, the parents were asked to complete a rather comprehensive inventory, and 98% of them complied.

DATA FROM THE INITIAL GROUP AT ADULT AGE

In the follow-up at adult age, the subjects of the initial group participated in a mail survey investigation at 23 years of age. In an inventory the subjects were asked about (a) educational and vocational activities from the end of the compulsory school at 16 up to the time of the investigation, (b) the current situation (1975) with regard to education and occupation, and (c) plans and expectations for the future with regard to education and vocation. The inventory was answered by 93% of those who had finished the gymnasium education (at the age of 19) and by 86% of those who left school at the age of 16. This investigation, among other things, functioned as a preparation to the follow-up of the main group at adult age.

DATA FROM THE MAIN GROUP AT ADULT AGE

A comprehensive follow-up was initiated for the main group when the subjects were 26 years of age. It consisted of (a) an investigation of the total group and (b) a complementary investigation of the biological sample.

Total Group Investigation

An inventory was administered to the total group. It covered the following aspects of the total life situation at the time of the investigation:

1. The present family situation
2. The present work situation
3. Education to the present stage
4. Vocation to the present stage
5. Further educational and vocational plans
6. Perceived possibilities to fulfill own plans and perceived obstacles
7. Leisure time activities
8. Social networks
9. Alcohol habits

The inventory was administered by mail to 1,358 males and females. It was answered by 85% of the total group, 80% of the males, and 90% of the females. A careful description of the group of respondents and an analysis of their representativeness has been performed (Andersson, Magnusson, & Dunér, 1983). Some characteristics are reported in Chapter 5.

For the evaluation of the importance of the drop out rate in the interpretation of the results of the empirical studies, it is important to observe that the dropout group was fully controlled in the sense that their characteristics were covered by various kinds of data for basic person factors as well as for environmental factors from the time at school, and by register data from official records. Some of these data, for example, teachers ratings and data from the records, are without any drop out and are thus very useful for characterization of the dropouts in the survey investigation.

Sample Investigation

Supplementary data that covered important aspects of psychological and biological functioning were collected at adult age for the biological sample when the subjects were 26–28 years of age. The subjects then participated in the following data collections: (a) An individual interview, (b) a test session, and (c) a medical examination. The number of subjects who participated in these data collections, the variables that were covered, the instruments used, and the procedures for data collections are reported in detail by Backenroth, Magnusson, and Dunér (1983) and Backenroth and Magnusson (1983). The main types of information that were collected in the three sessions are listed below.

1. Each subject participated in an interview session, which was aimed at deepening the information obtained from the answers to the inventory which had been administered by mail earlier. The interview lasted about 2-1/2 hours. It covered the following areas:

a) Work and education
b) Stressful events during the last six months
c) The use of drugs
d) Alcohol habits
e) Leisure-time activities
f) Experiences and memories of the subject's childhood and current relations with parents and siblings
g) Own family situation
h) Friends
i) Sense of control over own life
j) Working habits and possible stress reactions (Type-A behavior)
k) Occurrences of physical and mental illness
l) Important life events
m) Life values
n) Satisfaction with private life, with work situation, and with society.

Information about some of these areas was obtained by inventories which were answered during the interview session. This was the case for the aspects covered by j–n above.

Of the original sample, 230 subjects were available for contact and 198 of these agreed to participate in the interview session.

2. The *test session* took 2 to 2-1/2 hours. It included tests for the following factors:
   a) Intelligence
   b) Field dependence-independence (Rod and Frame Test)
   c) Masculinity-Femininity (Bem scale, 1974)
   d) Parent-Child relations (PCR)
   e) Memory (Memory for Design-MFD)
   f) Neurophysiological test battery (Halstead-Reitan):
      Abstract concept formation (Halstead Category Test -HCT)
      Survey flexibility (Trail Making Test - TMT)
   g) Personality (KSP): Psychic Anxiety; Somatic Anxiety; Muscle Tension; Social Desirability; Impulsivity; Monotony Avoidance; Detachment; Psychasthenia; Socialization; Indirect Aggressiveness; Verbal Aggressiveness; Irritability; Suspicion; Guilt; Inhibited Aggression.

Of the 230 subjects who could be reached, 171 participated in the test session. All of them had also participated in the interview.

3. The *medical* examination included the collection of the following material and data:
   a) Two measures of blood pressure under standardized conditions
   b) Blood sample and bleeding time
   c) Two laboratory samples of urine and a one-day production of urine
   d) Height and weight
   e) Data about fertility, medication, smoking habits, and the use of contraceptives.
      173 subjects participated in the medical examination.

DATA FROM OFFICIAL RECORDS FOR THE MAIN GROUP

The principal data in the project were obtained directly from the subjects themselves, and from the parents and teachers. These data have been described above. The information obtained directly is, of course, most valuable. However, important information about relevant factors needed for the elucidation of individuals' developmental process, cannot be obtained from these sources. Therefore, for the main group, the information obtained from the subjects, the parents, and the teachers was supplemented with information from other sources, particularly from official registers. Such data have played an important role in fulfilling the purpose of the project. They have been collected for two purposes: (a) to cover relevant aspects of the individuals and their environments that could not be covered in other ways, and (b) to complement with objective data the information about aspects that have been covered only with subjective data.

There are two important advantages in the collection and use of data from official records in Sweden. First, the Swedish registration system is very effective.

It has been possible to trace the places where each of the subjects have lived and to gain access to all official records; for example, those concerning registered criminal offenses and drug and alcohol abuse, as well as psychiatric records from all psychiatric clinics in the country, where the subjects have been diagnosed and treated. Second, the records are comparable as they are kept according to common rules across the country.

With due permission from the relevant legal authorities, data covering the following types of information were collected from national and local official sources: information from obstetric records (e.g., length and weight at birth, gestation time, and clinical evaluation of the new-born), information about the family structure of the participants during their childhood (e.g., one-two parent family, number of siblings, and geographical mobility), and information about alcohol abuse, criminal offenses, and psychiatric care. Reports of these data have been presented by Andersson and Magnusson (1986), Lagerström, Nyström, Bremme, Eneroth, and Magnusson (1985), Stattin, Magnusson, and Reichel (1986), and von Knorring, Andersson, and Magnusson (1987). The main characteristics of the participants of the main cohort with respect to alcohol problems, criminal offenses, and psychiatric care are given in Chapter 5.

For the study of the central problems in IDA, the records that were collected from official sources have advantageous characteristics. They reflect actual events that are important for the elucidation of the developmental process, in an objective way. They are relatively inexpensive to collect, they cover the entire life period, and the information that they contain is often recorded contemporaneously with the events (i.e., before later outcomes can influence the selection and interpretation of events to be recorded).

On the other hand, the problem connected with the use of such data is that in some areas, records are biased measures of the adjustment problems under consideration, because of the influence of other, irrelevant factors. Particularly sensitive in that respect are records for alcohol problems and criminal offenses (for a discussion of the problems connected with the use of records in research on delinquency and criminal offenses, see Farrington, 1979; Sparks, Genn, & Dodd, 1977). Two problems are connected with the use of data in these areas. First, only a fraction of individuals with alcohol problems and/or criminal offenses come into contact with the legal authorities and are registered. Second, the possibility of being recorded for alcohol abuse or criminal offense(s) can be dependent on other, person-bound factors such as intelligence, socio-economic situation, and education. Records in these areas are therefore of limited validity as bases for the estimation of the prevalence of alcohol problems and criminal activity in the total population.

The extent to which these limitations make these data less useful varies with the type of problem under consideration. In some cases, including those just discussed, the influence of such factors have been demonstrated empirically and must be addressed. However, in other areas, there is no reason to assume that irrelevant factors have had the same kind of systematic biasing effect on the data. This is the case, for example, with obstetric data and with data for family conditions during the upbringing of the children.

Comments

For the purpose of the longitudinal project, two characteristics of the data from official records are particularly important.

1. As mentioned above, the data are objective and reflect important events in the lives of the individuals.
2. The data from official records cover the total group without any drop out. This is an unusual and very favorable condition for the study of fundamental problems. Combined with data from an early age, that are representative and did not suffer from attrition (teachers' ratings, etc.), the data from official records permit control of effects on data for other factors.

Only a few longitudinal studies on the developmental background of adult maladjustment in terms of alcohol problems, criminal offenses, and psychiatric care have been performed and reported. Insofar as they have been carried out, they have been primarily concerned with only one or a few of these aspects. However, as has been demonstrated empirically, the same persons tend to be arrested as are divorced, placed in mental hospitals, become alcoholics, and display other problems (Farrington, 1979; Robins, 1966, 1978). Additionally, the same people tend to have been truants, thieves, aggressors, neurotics and problem children at school. Therefore important aspects of the total functioning of these individuals and the developmental process leading to that state of affairs are lost, if only one or a few aspects are studied at a time. The advantage of the total body of information that is available for the same persons from a variety of sources including official records, for a representative group of subjects who have been followed since they were ten, is obvious. Among other things, it permits the subjects to be grouped, at adult age, on the basis of their maladjustment patterns, and it permits the developmental background of these patterns to be investigated. In Chapter 8 an empirical study is reported that illustrates the usefulness of these data.

ETHICAL CONSIDERATIONS

One basic ethical aspect of psychological research in general and in a longitudinal project, in particular, has to do with the information to the participants and others concerned. As described in the previous section, considerable efforts have been made to take this aspect into account at all stages of the project.

Another important aspect has to do with the steps taken to guarantee that personal data are handled in such a way that confidentiality is ensured as far as possible.

When the first data collections were planned in the middle of the 1960s, the issue of protection of individuals' privacy had not become as heated as it is in many Western countries today. However, from the very start of the project, all possible measures were taken to protect the individuals from the possibility of being identified in the data, and to make sure that confidentiality was maintained throughout the data collection, data treatment and storage, and the presentation of results.

All data collections were performed by members of the research team. The teachers were not allowed to be present in the classroom during the data collections, so as to avoid influence on the pupils' willingness to respond and respond correctly. The instruments used were placed in prepared envelopes by the participants, who personally sealed the envelope before handing it over to the researcher. The pupils were promised that no information would be given to the school and that the sealed envelopes would not be opened until they reached the department in Stockholm. The instruments were constructed so that all identification of the individuals could be immediately removed and substituted with code numbers. All data were coded from the first punching of the cards in 1965. No researcher has been or is allowed to remove any data from the data base; all calculations have been and are performed by the project personnel and no individual data, only group data have been presented to anyone outside the project. Individual identity is not revealed in any report; only group characteristics are reported.

SCIENTIFIC COOPERATION

The planning and implementation of the project has by necessity implied comprehensive cooperation with researchers from many scientific disciplines, given the broad character of the project. This cooperation has included researchers from sociology, criminology, and medicine, particularly endocrinology, pharmacology, and neuropsychology. The researchers from these fields have been very cooperative and constructive, and their contributions to the planning and implementation of the project, the analyses of the data, and the interpretation of the results have been invaluable.

Part II

EMPIRICAL STUDIES

Four empirical studies are presented in which central issues are investigated, using data from the project. The studies are concerned with problems at various levels of complexity; they are performed as combinations of cross-sectional and longitudinal analyses and use variable-oriented regression models and pattern-oriented classification methods for treatment of data.

Chapter 5
CHARACTERISTICS OF THE MAIN COHORT

Now we can begin answering the central questions with which the project began. How has the life course progressed, and how does the general life situation appear for the main cohort at adulthood, at the time of the follow-up studies? In answering this question, data from the general follow-up inventory and from official records for the whole main cohort are presented in this chapter. The data cover the present situation at adulthood and the life course with respect to criminal offenses, psychiatric care, and alcohol problems.

GENERAL ASPECTS OF THE LIFE SITUATION AT AGE 26

As reported in Chapter 4, an inventory was administered to the main cohort when the participants were about 26 years of age. The purpose of the administration of the inventory was to cover important aspects of the life situation for the cohort, at an age when most have finished their education and a large part have established family life. The data were analyzed in detail and presented by Andersson, Magnusson and Dunér (1983). In this chapter, results are summarized for some main characteristics of the life situation for the participants, as reflected in their own reports at the age of 26.

Family Situation

In Table 5.1 some data for family life as reported at the age of 26 are given. Of those answering the inventory, 58% of the males and 71% of the females reported that they were married or living with someone of the opposite sex. Twen-

TABLE 5.1.
Family Situation for Males and Females at Age 26 (Percentages).

		Males	Females
Married/living together		58	71
Children	0	71	54
	1	20	28
	2	8	15
	3–5	1	3

ty-nine percent of the males and 46% of the females had children. An interesting observation was that children are much more commonly reported by single females than by single males. Almost one out of five (19%) of the females living alone had children, while this was reported by only 7% of the males.

Education and Work

The level of completed education and the present situation with respect to education and work are presented in Table 5.2.

The compulsory school system in Sweden covers nine years, from the age of 7 to the age of 16. More than 80% of those answering the inventory continued with some form of education after compulsory school. Among those who proceeded to higher theoretical education, there is a clear sex difference. To a marked extent, more males than females chose further, theoretical education.

In total, 3 out of 10 reported that they thought about further education at the age of 26. For most of the participants the plans seemed to be independent of

TABLE 5.2.
Education and Work Situation at Age 26 (Percentages).

	Males	Females
Completed education:		
Theoretical education beyond compulsory level	46	53
Professional training beyond compulsory level	38	31
Compulsory school only	16	16
Present situation:		
Full time work	82	46
Part time work	4	26
Full time education	12	12
Part time education	4	4

external factors. Sixty-nine percent saw no obstacles to going into further education of some kind.

A rather traditional gender difference was revealed with respect to work and type of work. Most of the males were working full time and more than half of them within technical industry and skilled crafts professions. Only 46% of the females were working full time and most of them in caretaking or administrative professions. Of the females, 26% were working part time, while this was the case for only 4% of the males. Of those who did not work or study, 38 females and 2 males reported that they had home care as their main occupation.

Ninety percent reported that they liked their present work situation, and 74% reported that they had chosen the right type of work. Though both males and females were satisfied to the same extent, more males (22%) than females (16%) thought about changing their job position.

On the average, both males and females had been working for 7 years. Generally, both males and females had tried various types of work. Eighteen percent had stayed at their first job, and 24% had had five jobs or more. About one third of all 26-year-olds had been unemployed at least once. The quality of the working environment and job security were the factors that were most highly rated among both males and females. Next in importance, females preferred jobs in which they could care for and have contact with others, while males emphasized salary as a factor in job choice.

Sense of Control Over Own Life Situation

An important aspect of an individual's life situation is the experienced capability of influencing and controlling one's life situation by his or her own actions. This aspect can be regarded as an important factor for intrinsic adjustment. In Table 5.3 the answers to two questions concerned with this issue are reported.

Most of the participants reported that they felt they could influence and control the present life situation to a large extent. Less than 10% of both sexes

TABLE 5.3.
The Experienced Sense of Control of Own Life Situation (Percentages).

	The Present Situation		*The Future*	
	Males	*Females*	*Males*	*Females*
To a large extent	81	84	69	72
To some extent	11	11	18	18
To a low extent	9	6	12	10

TABLE 5.4.
Leisure-Time Activities (Percentages).

Question	*Males*	*Females*
Can you use your spare time satisfactorily?:		
Most of the time	51	53
Often	25	26
Sometimes	15	13
Not so often	6	7
Almost never	2	1
Activities during spare time[1]:		
Sports	49	31
Reading	65	72
Religion	5	4
TV-watching	71	74
Studying	25	22
Dancing	9	13

[1]At least two hours/week

experienced this possibility as low. When the question was posed concerning the future, the experienced sense of control and possibility to influence was lower for both sexes. It is interesting to note that no marked sex difference appears in this important aspect of the total life situation.

Leisure-Time Activities

In Table 5.4 the answers to some questions concerned with the participants' use of their leisure time are reported.

Most males and females were satisfied with the extent to which they could use their leisure time. The most common activities were watching TV, reading, and sports. Two out of three reported that they belonged to a club, association, or society. Forty-seven percent of these males and 31% of these females belonged to a sports club, but only a few to a religious (10% of the males and 9% of the females) or political organization (7% of both sexes). While 35% of the males devoted more than 8 hours a week to club activities, mainly sports, this was the case for only 15% of the females.

Social Relations

A number of inventory questions dealt with the participants' social network. Data for some aspects of relations to various others are presented in Table 5.5.

Almost all of the participants frequently saw someone. Even qualitatively the relations to others were good for most of both males and females. About 90% expected full support from their parents if they were in a troublesome situation.

TABLE 5.5
Social Relations Excluding Relations with Husbands, Wives, Cohabitants (Percentages).

	Frequent Contact		*Discusses Problems*		*Support Available When Needed*	
	M	*F*	*M*	*F*	*M*	*F*
Parents	75	81	64	70	88	91
Siblings	59	63	43	48	62	59
Relatives	22	33	7	10	22	25
Friends, acquaintances	98	97	63	82	64	62
No one	1	0	11	3	3	1

On the whole, relations with parents seemed to be good. About two out of three could talk openly with the parents about their problems. More females than males had someone to confide in, and more often females had several others with whom they could talk intimately.

Comment

The areas covered by the inventory make up a limited part of the total life sphere of the participants. Therefore, it is not possible to draw more general conclusions about their total life situation. The fact that the majority have some common, positive characteristics cannot be used for an evaluation of the life situation for single individuals. It should also be remembered that 10% of the females and 20% of the males did not answer the inventory. An analysis of the drop-out group using the data available from the school period does not, however, reveal any dramatic difference between the dropout group and the respondents with respect to adjustment at an earlier age (see Andersson, Magnusson, & Dunér, 1983).

CRIMINAL ACTIVITY

As reported in Chapter 4, information about committed crimes was collected from all official local and national registers. This implies that data for criminal acts are available for all individuals in the main cohort. The full procedure for the data collection and the detailed analysis of data for criminal acts and sentences were reported by Stattin, Magnusson, and Reichel (1986). Characteristics of the main group with respect to age at first conviction and prevalence of crime at different age stages and for different types of crimes are summarized in the following pages.

Data for criminal acts are available for all individuals from birth up to and including age 29. A crime was defined using the following criteria: For an adult, a court decision of guilty would determine if the act was to be regarded as a crime. For children below the age of criminal responsibility (15 years), a crime was counted if the individual, after a local hearing of the local police, was handed over to child welfare authorities.

Age at First Conviction

To better understand the developmental process underlying a criminal career, the age at which the first crime was committed is of interest. The figures showing this for males and females are given in Table 5.6.

At the age of 30, 267 males or 37.7% of the total group had been registered for a crime. The mean age for the first conviction was 17 years, 2 months and the median age 16 years, 7 months. As shown in Table 5.6, the first crime is generally committed during the period of 15–17 years. A closer inspection of the figures for each year reveals that the first crime occurs during the ages of 14, 15, and 16. The figures demonstrate the importance of collecting data from the local police on criminal offenses at an early age. At age 15, when the boys reached the age of criminal responsibility and could be included in the National Police Board register, about 30% of those with registered offenses had already been registered by the local police and the local social authorities.

Of those who had committed a crime before the age of 30, about three out of four males had committed their first crime before the age of 21. Less than 4% had been convicted for the first crime after the age of 27. However, these figures imply that about one out of four makes his criminal debut after the age of 21.

At the age of 30, 61 females or 9% of the total group had been registered for a crime. The mean age for the debut was 22 years, 3 months and the median age

TABLE 5.6.
Age at First Conviction for Males and Females
in the Main Cohort.

	Males		*Females*	
Age	*N*	%	*N*	%
Up to 11	31	11.6	2	3.3
12–14	50	18.7	7	11.5
15–17	83	31.1	6	9.8
18–20	38	14.2	14	23.0
21–23	32	12.0	20	32.8
24–26	23	8.6	6	9.8
27–29	10	3.7	6	9.8
Total	267	100.0	61	100.0

TABLE 5.7.
Number and Percentage of Males and Females Registered at Different Age Periods.

	Males		Females	
Age	*N*	%	*N*	%
up to 11	31	4.4	2	0.3
12–14	66	9.3	8	1.2
15–17	133	18.8	10	1.5
18–20	92	13.0	20	2.9
21–23	100	14.1	27	4.0
24–26	68	9.6	13	1.9
27–29	54	7.6	10	1.5
Aggregate data				
Youth (<18)	165	23.3	15	2.2
Early adulthood (18–29)	190	26.8	58	8.5
Total period (up to 29)	267	37.7	61	9.0

21 years, 2 months. As can be seen from Table 5.6, females engage in criminal activities far less often and substantially later than males. More than half of the females were registered for the first time during the ages of 18 through 23, typically between the ages of 21 through 23.

Criminal Activity at Different Ages

Questions about the age during which males and females are most criminally active are important but cannot be unambiguously answered. The answer may vary with the way in which criminal activity is measured, that is, if it measures the number of individuals who are active, the number of crimes that are counted or the number of crime occasions at each age level. In Table 5.7 the figures are given for the number of individuals who were registered in official records at various age levels.

The highest proportion of males being registered for crimes is found during the age period 15–17. In a broader perspective, the peak period of criminal activity is during the period 15–23. The same general picture emerges when the number of registered crime occasions is used as a measure. A total of 1,841 crime occasions were registered for the males before the age of 30, or 6.9 crime occasions per male individual. Defined as the number of crime occasions per individual, the individual activity was lowest before the age of criminal responsibility, reached a peak between ages 15 and 21, declined somewhat during the following six years, and reached a second peak during the last age period (Stattin, Magnusson, & Reichel, 1986).

For females, there are only small variations across age periods in the number of individuals who have been registered for crimes. Like males, there is a tendency toward higher activity in the total group during the age periods at which the criminal activity starts, that is, for females during the period of 18 through 23. This trend holds also when criminal activity is studied in terms of the number of crime occasions. There is no clear age related tendency for females regarding registered number of crime occasions per person.

Types of Crimes at Different Ages

In descending order, the most frequent crimes for males (expressed by the percentage of individuals who have been registered for each crime) were the following: unlawful possession of property (13.3%), theft (9.9%), unlawful driving (9.7%), grand larceny (7.2%), petty theft (7.2%), property damage (7%), driving under the influence of alcohol (6.9%), unlawful taking of a vehicle (6.5%), assault (5.8%), and careless driving (5.4%). For females the most frequent crimes, expressed in the same way, were: unlawful possession of property (3.2%), petty theft (2.3%), driving under the influence of alcohol (1.2%), careless driving (0.9%), fraud (0.9%), and grand larceny (0.9%).

In the preceding sections those who have been registered have been dealt with as one homogeneous group, independent of the character of the offenses. In order to study more systematically how the type of crimes varies with the age level, all offenses were classified into the following categories: Violent offenses against persons, Property damage, Crimes for personal gain, Traffic offenses (alcohol and non-alcohol), Narcotics offenses, and Miscellaneous offenses (Stattin et al., 1986). In Table 5.8 the percentages of the registered males involved in these types of crime at different ages are reported.

The figures in table 5.8 are based on this differentiation. For each age are given the percentages of those registered who have committed various types of crime. It should be observed that each individual can have been registered for more than one crime on the same occasion. The number of registered crimes may be somewhat biased for the early ages, due to the fact that the local police report only major offenses for boys under 15. This is the reason why the last row in Table 5.8 includes only individuals from the age of 15.

As can be seen in Table 5.8, the highest occurrence rate occurs for Crimes for personal gain (property crime) for all age periods. It is somewhat more salient in the younger than in the older males. Of the registered males, 71.5% had committed one or several crimes of this type between the ages of 15 and 30. The second most common offense is Traffic offenses of the nonalcohol type; they begin in the age period 15 through 17 years. About 4 out of 10 registered boys between 15 and 30 years were registered for such a crime. Traffic offenses involving alcohol are also common. Violent offenses against other persons rarely occur before age 15. Altogether, 61 males (8.6% of all males and 24.5% of the

TABLE 5.8.
Type of Crime Committed by Males at Different Ages (Percentages of Those Registered).

Age	*Violent Offenses Against Persons*	*Property Damage*	*Crimes for Personal Gain*	*Traffic Offenses: Alcohol*	*Traffic Offenses: Non-Alcohol*	*Narcotics Offenses*	*Miscel-laneous Offenses*
up to 11	0	38.7	67.7	0	6.5	0	0
12–14	9.1	18.2	83.3	0	9.1	0	0
15–17	18.0	7.5	69.2	11.3	39.8	2.3	9.0
18–20	23.9	14.1	69.6	23.9	40.2	5.4	6.5
21–23	18.0	14.0	63.0	22.0	27.0	5.0	4.0
24–26	23.5	20.6	47.1	23.5	41.2	8.8	8.8
27–29	27.8	20.4	53.7	20.4	31.5	18.5	9.3
15–29	24.5	17.3	71.5	25.7	42.6	9.2	10.0

registered males) had committed such a crime by age 30. The number of males committing Property damage was evenly distributed from an early age up to age 26 but was slightly more prevalent among the youngest males. Narcotics offenses were relatively rare. However, 18.5% of the registered males in the age period 27 through 29 years had committed an offense against the Narcotic Drug Act.

Thus, before the age of criminal responsibility (age 15), males with official delinquency records generally commit crimes for personal gain and property damage. A wider distribution of crimes then occurs with an increase of alcohol- and nonalcohol-related traffic offenses and of violent crimes.

In Table 5.9 the number of females registered for the various types of crimes is reported for each age level. The figures are based on the number of recorded crimes per person independent of how many crime occasions were registered.

The dominant type of crime among females, Crime for personal gain, was the same as it was for males. Seven out of ten registered girls were recorded for this crime category before age 30. The wide distribution across crime types common for boys was less prevalent for girls. Although about the same percentages of registered females committed Traffic offenses involving alcohol, a considerably lower percentage of females committed Narcotics offenses, Traffic offenses of the non-alcohol type, Property damage, and Violent offenses.

Registered Criminality in a Longitudinal Perspective

The figures presented so far do not provide information about the course of criminality of individuals, that is, if and to what extent it is the same individuals who commit crimes at different age levels. Surprisingly few empirical results have been presented in the criminological literature on this important issue. Such an investigation requires linking information on offenses committed at

TABLE 5.9.
Types of Crimes Committed by Females at Different Ages.
(Frequencies)

Age	*Violent Offenses*	*Property Damage*	*Crime for Personal Gain*	*Traffic Offenses: Alcohol*	*Traffic Offenses: Nonalcohol*	*Narcotic Offenses*	*Miscellaneous Offenses*
up to 11		1	1				
12–14	1		8				
15–17	2	1	10	1	1		1
18–20	1		12	3	5		1
21–23	1	1	20	3	3	2	1
24–26	1		6	4	2		2
27–29			7	2	1		1
15–29	6	2	40	12	11	2	6

different ages on an individual basis. It is this link that enables us to observe when criminal behavior starts, when it peaks, and when it ends for the particular individuals under investigation. It enables us to study the different types of individual criminal careers of such persons who are criminally active recurrently from an early age to adulthood, of those who commit a large number of offenses during a short period in their youth, of those who enter the criminal scene late in life, of those who tend to specialize in a certain type of crime, and of those who show an indifferentiated offense pattern across time.

Because detailed information has been collected about the age level at which each offense has been registered and about the specific type of crime that has been committed on each occasion, further analyses concerned with the questions raised above are possible and under way in the project. In Table 5.10, an overview of criminal activity in a longitudinal perspective is given for males.

Table 5.10 is rich with information. However, even more information is contained in such a presentation when data refer to specific types of crime. Analyses of that kind are under preparation. Here, only a few trends of particular interest will be emphasized. The most typical outcome at adulthood is for those who were registered at both the earlier stages and for those who were not registered at all during these periods. For the first group, 38 out of 55, or 69.1%, were also registered at adulthood, while the figures for the other group were 64 out of 506, or 12.6%. These two groups constitute significant types in a configurational frequency analysis. In other words, they occur significantly more often than expected if the individuals were distributed randomly in the final categories. It is

TABLE 5.10.
Criminal Activity Among Males in a Longitudinal Perspective

Childhood (0–14)	*Adolescence (15–19)*	*Early Adulthood (20–29)*	*Obtained*		*Expected*
			N	%	*N*
Criminal (N = 81)	Criminal (N = 55)	Criminal	38[t]	69.1	4
		Noncriminal	17		16
	Noncriminal (N = 26)	Criminal	8	30.7	12
		Noncriminal	18[at]		47
Noncriminal (N = 628)	Criminal (N = 122)	Criminal	41	33.6	33
		Noncriminal	81[at]		124
	Noncriminal (N = 506)	Criminal	64[at]	12.6	100
		Noncriminal	442[t]		371

[t] = type significant at the 5 percent level
[at] = antitype significant at the 5 percent level

interesting to note that three significant antitypes occur: those who were registered only before 15, those who were registered only during adolescence, and those who were registered only at adulthood. These characteristics appear significantly less often than could be expected from a random model.

Comment

Because national crime registers in Sweden do not contain information about crimes committed by minors under the age of legal responsibility, data cannot be established as correctly as for older ages. The exclusion of young persons from official statistics on crime in Sweden has posed problems for international comparisons, and has been considered a main reason that prevalence figures for the Scandinavian countries generally have been lower than those from the rest of Europe. The collection of data by our research group on the local level for under-age crime offers somewhat better possibilities for making accurate comparisons. Table 5.11 reports prevalence figures for non-traffic offenses for two Swedish cohorts and a nationwide population in England and Wales based on official statistics.

As can be seen in Table 5.11, the prevalence figures at the different ages are almost identical for the two Swedish cohorts and differ very little from the prevalence estimates for British citizens.

To summarize, our data show that it is not a rare event for a boy to be registered by the police. In the present study, more than every third male (37.7%) was registered for one or more criminal offense(s) up to age 30. This prevalence figure includes traffic incidents. However, excluding all types of traffic offenses (alcohol and nonalcohol related), 32.5% of the males still would be found registered before age 30. It should be noted that the offenses drunkenness and disorderly conduct, are not included in the above figures. With a definition of crime as the total number of registered police contacts leading to arrest or temporary custody, court convictions, or treatment by child welfare

TABLE 5.11.
Prevalence of Crime at Different Ages for Swedish and British Populations (Percentages).

Prevalence Before	*The Present Study (N = 709)*	*Jansson 1981 (N = 7719)*	*Farrington 1981a (N = 9000)*
15	11.1	10.3	11.7
18	20.2	19.1	—
21	24.5	—	21.8
26	31.2	31.0	—
30	32.5	—	30.3

authorities, 40.8% of all males in our research group were recorded for at least one offense before the age of 30.

The prevalence figures for females are lower. Nine percent had committed one or more offense(s) before age 30. Excluding the traffic offenses, 7.2% of the females had been registered. Finally, including the females with drunkenness offenses or disorderly conduct beyond the 9% who had court convictions or early contacts with child welfare authorities for a law-breaking act, 10.3% of all females in the research group had at least one police contact before the age of 30.

PSYCHIATRIC CARE

Information about psychiatric care was obtained for inpatient as well as outpatient treatment at all psychiatric clinics in Sweden. These data are almost complete for the cohort. The information covers the age period 0–24. The psychiatric records were classified by a trained psychiatrist according to DSM III, and a sample of the records was reclassified by an independent psychiatrist to estimate the reliability of the classification. The procedures for data collections and for the analyses of the records, have been presented by von Knorring, Andersson, and Magnusson (1987).

Before the age of 25, the cumulative percentages of males and females who had been registered for psychiatric care were 13.9 and 14.2, respectively. Here the incidence and prevalence of psychiatric care will be presented with reference to age level, together with some information about the developmental course of psychiatric disorders.

Incidence of Psychiatric Disorder

The incidence of psychiatric disorders for the males and females is shown in Fig. 5.1.

The most common first diagnosis on axis I in DSM III was anxiety disorders, for both males and females. Anxiety disorders start from 5 years of age, but the most common age of onset was during early adulthood (20–24 years of age). The second most common first diagnosis for males was substance abuse disorders. For females, the second most common diagnostic category was affective disorders and adjustment disorders with depressed mood. There is a conspicuous gender difference in the trend over age. As can be seen from Fig. 5.1, the incidence of psychiatric disorders is low during the first five years of life. The incidence slowly and continuously increases with age. During childhood, boys are more commonly referred to a psychiatrist than girls. During early adolescence the figures are about equal for the sexes, and during late adolescence there are about twice as many females consulting a psychiatrist for the first time.

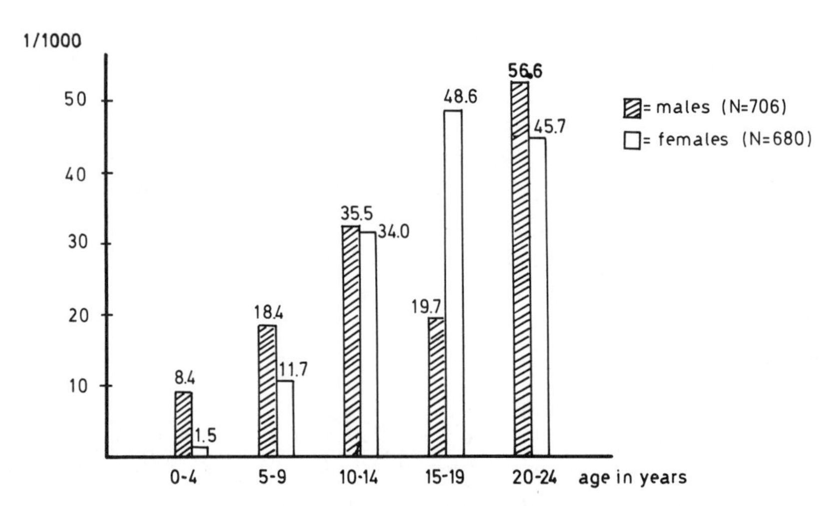

FIGURE 5.1. 5-year incidence of psychiatric disorders - first time consulting a psychiatrist. Diagnosis according to axis I + II, DSM-III.

Prevalence of Psychiatric Care

In Table 5.12 the figures for prevalence of psychiatric disorders for various age levels are given for males and females.

As can be seen in Table 5.12, the prevalence of psychiatric disorders increases with age. During the first 10 years, no diagnosis appears with a high frequency, neither for boys nor for girls. During the age period 10–14, attention deficit disorders and conduct disorders were the most frequent diagnoses for both sexes. They remained the most frequent ones for males in late adolescence but disappeared totally in early male adulthood. Instead, anxiety disorders and non-depressed adjustment disorders, together with substance abuse disorders, appeared most frequently during this period. For females, both late adolescence and early adulthood were characterized by relatively high frequencies of substance abuse disorders and adjustment disorders with depressed mood and by anxiety disorders and non-depressed adjustment disorders.

The Development of Psychiatric Disorders

The prognosis for participants with different ages at the onset of psychiatric disorders is shown in Figure 5.2. For the children who had psychiatric disorders beginning before 10 years of age, the outcome is quite good. Only 11.0% of the boys and 11.4% of the girls remained in psychiatric care as young adults. Almost one third (31.9%) of the males with onset of psychiatric disorders at 10–14 years were still in psychiatric care as young adults. Less than every fifth female (17.4%) with onset of psychiatric disorder at 10–14 years of age was still in psychiatric

TABLE 5.12.
5-Year Period Frequencies for Prevalence of Psychiatric Disorders
During Different Age Intervals.
Males N = 706, and Females N = 680. (Only the First Diagnosis Given During Each Interval Was Counted.)

	Males						*Females*					
	0–4	*5–9*	*10–14*	*15–19*	*20–24*	*0–24*	*0–4*	*5–9*	*10–14*	*15–19*	*20–24*	*0–24*
1. Mental retardation and Organic mental disorders	1	1	—	1	1	3	—	—	2	3	1	3
2. Attention deficit disorders and conduct disorders	—	4	16	11	—	18	—	2	11	6	—	13
3. Somatoform disorders, other disorders with physical manifestations, stereotyped movement disorders, and eating disorders	1	5	7	1	1	11	1	3	3	3	3	10
4. Substance abuse disorders	—	—	3	9	17	14	—	—	2	12	11	12
5. Schizophrenic disorders and other nonaffective psychoses	—	—	1	1	1	1	—	1	—	1	1	2
6. Affective disorders and adjustment disorders with depressed mood	—	—	1	1	8	8	—	1	1	12	16	22
7. Anxiety disorders and adjustment disorders, nondepressed	1	4	5	4	29	33	—	2	8	10	19	31
8. Other disorders	2	4	—	3	2	5	—	—	—	1	1	2
Axis II	1	1	1	1	3	5	—	—	—	—	1	1
Total	6	19	34	32	62	98	1	9	27	48	53	96
Suicide	—	—	—	1	—	1	—	—	—	—	1	1

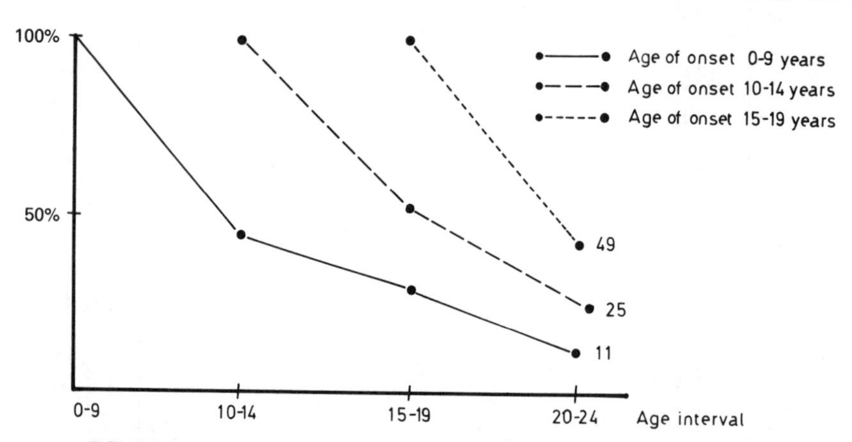

FIGURE 5.2. Longitudinal course of psychiatric disorders for different ages of onset. Percentage still being in psychiatric care at different ages.

care during early adulthood. About a third of the males having onset of psychiatric disorders in late adolescence remained in psychiatric care as young adults. Slightly fewer than half of the women with onset in late adolescence were in psychiatric care as young adults.

About 40% of those who had conduct disorders at an early age later manifested substance use disorders (41.7% males and 38.5% females). In most cases, substance abuse was of alcohol (81.5%); only a minority abused drugs exclusively.

Comment

Figures for the incidence and prevalence of psychiatric disorders at various age levels, as well as figures for the expected long-term outcomes have been presented for various samples of males and females (e.g., Baldwin, 1968; Blanchard & Paynter, 1924; Cederblad, 1983; Graham & Rutter, 1976; Jonsson & Kälvesten, 1964; Lavik, 1976; McGulloc, Henderson, & Philip, 1966; Robins, 1966; Rutter, 1972; Rutter & Graham, 1968; Rutter et al., 1970, 1975). The figures presented in these studies vary, sometimes widely. One reason is the variation in criteria that were used for the diagnoses. In order to interpret the figures that have been presented above in terms of registered psychiatric care, it should be remembered that the cohort studied included all those who attended the regular school at the age of 10, while those who could not, because of severe physical or mental handicaps, were excluded.

The figures presented for incidence and prevalence should be evaluated against the background of the rather low frequencies that occur in each cell of the tables. For example, the 90% confidence interval for the figure 31.9% for males is 17 to 47%. This fact means that all trends should be interpreted with caution.

ALCOHOL PROBLEMS

Data for alcohol use and abuse have been collected through self-reports by the subjects and from official records. The first type was collected by inventories administered at the ages of 14, 15, and 26, and by interview in the follow-up at age 26. Data from official records were collected from all possible sources in Sweden and covered the age period from 15 through 24 years of age. An analysis of the types of data available for alcohol have been reported by Andersson and Magnusson (1986). The primary characteristics of the main group with respect to alcohol abuse as reflected in the official records are reported below.

Registered Alcohol Abuse

The alcohol abuse registered for males and females is reported in Table 5.13 for the two age levels 15–17 and 18–24, respectively.

Of the main cohort, 7% of the males were registered for alcohol abuse at the age level of 15–17 and 14.5% at the age level of 18–24. Over the total age period, 17.4% of the males were registered before the age of 25, and 8.4% were registered on more than one occasion.

As expected, the number of females registered for alcohol abuse was small. Table 5.13 shows that 2.2% of the females appeared in the registers for alcohol abuse between 15 and 18 and 4% between 18 and 25. At the age of 25, 5.1% of the females had been registered at least once and 2.7% on more than one occasion.

Alcohol Abuse in a Longitudinal Perspective

In Table 5.14 data for registered alcohol abuse in a longitudinal perspective are reported for males.

TABLE 5.13.
Number and Percentages of Males and Females Registered for Alcohol Abuse.

	Males		*Females*	
Age	*Once*	*More Than Once*	*Once*	*More Than Once*
15–17	30 (4.3)	19 (2.7)	7 (1.0)	8 (1.2)
18–24	55 (7.9)	46 (6.6)	13 (1.9)	14 (2.1)
15–24	63 (9.0)	59 (8.4)	16 (2.4)	18 (2.7)

TABLE 5.14.
Registered Alcohol Abuse for Males in a Longitudinal Perspective.

Registered at 15–17	*N*	*Registered at 18–24 (Percentages)* Never	Once	More Than Once
Never	654	88.9	7.0	4.1
Once	30	56.7	20.0	23.3
More than once	19	21.1	15.8	63.2

Chi^2 = 132,5, df = 4 $p < .0001$

A total of 49 males were registered for alcohol abuse during the period 15 to 17. The probability is rather high that they also will appear in the registers between the ages 18 and 25. Among those who were registered once during the early period, 43.3% appear again during the later period. Among those who were registered more than once between 15 and 18, 79.0% are found in the registers for the period 18 to 25, 63.2% more than once.

The alcohol figures for females in a longitudinal perspective are given in Table 5.15.

For females as for males the probability is high that those who have been registered between the ages of 15 and 18 will appear in the registers again. Among the females who were registered before 18, 8 were also registered during the period of 18 to 25, 7 of them more than once. By comparison, only 2.9% of the girls who had not been registered at 15 to 18 were registered later.

Comments

Most youngsters enter the teenage period without experience with alcohol and leave it with well-established drinking habits (Wechsler, 1979). After a debut that most often appears in the beginning or the middle of the teenage years, alcohol

TABLE 5.15.
Registered Alcohol Abuse Among Females in a Longitudinal Perspective.

Registered 15–17	*N*	*Registered 18–24 (Percentages)* Never	Once	More Than Once
Never	652	97.1	1.8	1.1
Once	7	57.1	14.3	28.6
More than once	8	37.5	0.0	62.5

consumption increases rapidly with age; the period around 20 is usually described as the most intensive (Blane, 1979; Hibell, 1977; Zucker, 1979).

One reason for taking early drinking habits seriously is the assumption that youngsters who develop such habits early run a considerable risk of later alcohol abuse. Empirically, this important assumption has been inadequately investigated, and in particular there is a lack of research based on representative samples followed across time. Those empirical studies that do focus on the development of drinking habits and alcohol abuse across time are most often based on special groups such as young, first-time offenders (Collet, 1963; Eklund & Nylander, 1965; Nylander & Rydelius, 1973; Rydelius, 1978) or start too late to illuminate the role of early, initial drinking habits (Cahalan & Room, 1974; Fillmore, 1974; Loper, Kammier, & Hoffman, 1973).

The empirical study that comes closest to meeting the necessary requirements is the one presented by Donovan, Jessor, and Jessor (1983). They report the continuity in "problem drinking" over a 7-year period (17–24 years of age) based on self-reported data. According to the authors, the results do not indicate any strong stability in problem drinking. More than half of the boys who were regarded as problem drinkers early on were later regarded as nonproblem drinkers. The corresponding figure for the girls was 70%. Stability in nonproblem drinking was considerably stronger. In considering the results presented above from IDA and those presented by Donovan, Jessor, and Jessor, the difference in type of data should be observed; IDA used register data from official records, while the data used by Donovan, Jessor, and Jessor were self-reports. And, as the authors themselves noted, the generalizability of the Donovan, Jessor, and Jessor study is restricted by the rather large and possibly selective drop out of participants that occurred.

TO WHAT EXTENT ARE CRIMINALITY, PSYCHIATRIC CARE, AND ALCOHOL ABUSE ISOLATED PHENOMENA?

In the early sections, the characteristics of the main group have been presented for each of the serious maladjustment indicators (criminality, psychiatric illness, and alcohol abuse) as they are found in official records. With reference to the theoretical framework and its methodological implications described in Chapters 1–3, a central question concerns the extent to which these aspects of individual functioning can be seen as isolated phenomena as against their occurrence together in various patterns of inter-relationship. The answer to that question has important consequences for research strategies in these fields as well as for theories about the developmental process underlying adult maladjustment. This issue is addressed here in two ways.

TABLE 5.16.
Longitudinal Homotypic and Heterotypic Stability in Official Records for Criminal Offenses, Psychiatric Care, and Alcohol Abuse (Frequencies)

		Age 18–23					
		Criminality		*Psychiatric Care*		*Alcohol Abuse*	
		No	*Yes*	*No*	*Yes*	*No*	*Yes*
Age 0–17							
	No	471	70	509	32	508	33
Criminality		***		***		***	
	Yes	86	78	136	28	105	59
	No	526	123	598	51	576	73
Psychiatric care		***		n s		***	
	Yes	31	25	47	9	37	19
	No	540	116	609	47	589	67
Alcohol abuse		***		***		***	
	Yes	17	32	36	13	24	25

***$p < .001$ (Chi^2 test corrected for continuity with df=1)

In Table 5.16 the longitudinal, homotypic, and heterotypic relationships, from age 0–17 to age 18–23, are presented for the three types of maladjustment, measured by official record data.

A strong stability over time is obtained for criminality and for alcohol abuse, two aspects of maladjustment for which there exists a strong interdependence. The percentage of males with criminality records at age 18–23 is even higher for those with early alcohol abuse than for those with early criminality records (65% as compared to 48%).

Both criminality and alcohol abuse at age 18–23 are strongly related to psychiatric problems at age 0–17. Of those who had psychiatric records at an earlier age, 45% had criminal records at age 18–23, compared to 19% of those who had no psychiatric records at the earlier age. For those with alcohol records at age 18–23, the corresponding figures were 33 and 11%.

The longitudinal analysis of psychiatric data reveals an interesting pattern of relationships over time. Though there is a tendency in the expected direction, psychiatric care at the age of 18–23 is not significantly related to psychiatric care at an earlier age. However, it is significantly related to records at age 0–17 for both criminality and alcohol abuse. Of those with a criminal record at the earlier age, 17% had a psychiatric record at the later age period, while this was the case for only 6% of those who had no earlier criminal record. The corresponding figures for alcohol records were 26% and 7%, respectively.

Another aspect of the issue regarding the interdependence of various aspects of maladjustment is illuminated in Table 5.17. The frequencies with which criminality, psychiatric care, and alcohol abuse have been registered alone and in combination with each other are given for males in the main cohort. In order to avoid data contamination, those who were recorded in registers for criminal offenses and in registers for psychiatric care due to alcohol abuse were removed from those registers before the calculations presented in Tables 5.16 and 5.17.

At the age of 0–17, a total of 164 males were registered for criminal offenses. In 109 cases, criminality appeared alone. In 55 cases (33.5%), criminal records appeared together with records for psychiatric care and/or records for alcohol abuse. At the age of 18–23, the corresponding figures were 148 and 77 (52.0%). Thus, the percentage of males who have developed from criminal offenses as the only maladjustment indicator to a multiproblem syndrome increases dramatically.

Psychiatric care was registered for 56 males at age 0 to 17. In 27 of these cases (48.2%) psychiatric records appeared together with other records. At the age period 18 through 23, the corresponding figures were 60 and 35 (58.3%).

For alcohol abuse, 49 males were registered before 18. In 37 of these cases (75.5%), alcohol abuse was combined with records for criminal offenses and/or psychiatric care. The corresponding figures for the age level 18 through 23 were 92 and 70 (76.6%).

Some other interesting tendencies emerge in Table 5.17. The first is that alcohol problems appear together with other maladjustment indicators from the

TABLE 5.17.
The Frequency with which Criminality, Psychiatric Care, and Alcohol Abuse are Registered Alone and in Combination with Each Other in Males.

	Age	
	0–17	*18–23*
Criminality only	109	71
Psychiatric care only	29	25
Alcohol only	12	22
Crim+Psych	18	13
Crim+Alc	28	48
Psych+Alc	0	6
Crim+Psych+Alc	9	16
Total	205	201

beginning. About three out of four males have other records at both age levels if they have been registered for alcohol.

Conclusions

In this chapter a description has been given of the main cohort with respect to some characteristics that are of importance in fulfilling the aims of the longitudinal project. A major advantage to the description in these important respects is that it is based on data for a representative sample. The description of the cohort with respect to criminality, psychiatric care, and alcohol abuse is based on data that are as complete as possible. The drop out group that occurred for inventory data, which were used for the description of other aspects of the total life situation, does not invalidate the results to any substantial degree.

For those who answered the inventory, the overall picture was that at 26 years of age they were satisfied with their present situation. Most of them were satisfied with their work conditions and with their leisure-time opportunities, and they had good relations with others. Few of them saw serious obstacles to further education, if they wanted to continue, and felt that they had a reasonable degree of control over their life situations, particularly in the present. There was a general tendency toward a traditional sex role pattern. The females established themselves in a family situation earlier than the males; they worked in traditional female professions; and they had a greater number of other people to confide in.

As emphasized in Chapter 3, developmental research has been dominated by an interest in individual variables, that is, in single aspects of individual functioning. If this is true for research on normal development, it holds as well for research on maladjustment. Most of the time, serious forms of maladjustment such as criminality, psychiatric illness, and alcohol abuse are investigated separately. The fragmentation of research in these fields is reflected in the specialization among researchers into those concerned with psychiatric issues, criminological issues, and alcohol problems, as well as in the way research in these areas is published.

The results here presented for criminality, psychiatric illness, and alcohol abuse, as indicated by official records, clearly demonstrate the limitations imposed by this fragmentation on the traditional approach.

Of those having a record for alcohol abuse, about three out of four also have other records, and of those who have records for criminal offenses or psychiatric care, more than half also have at least one other record. Thus, it is more common for a person to have more than one problem. The important general conclusion that can be drawn from these figures is that effective research on each of the three aspects of maladjustment—criminality, psychiatric illness, and alcohol abuse—cannot be planned and implemented in isolation from the other, neither in a current nor in a developmental perspective. The separation of one group of individuals who, for example, have been registered for alcohol abuse without any other registration, from the group that has been registered for alcohol abuse *and* for psychiatric care and/or criminality contains important information that can be of decisive value for understanding the background of alcoholism. These two groups of individuals having been registered for alcohol abuse should be investigated separately. It would then be a mistake to assume that only those who are registered for alcohol abuse are the true alcoholics. Rather, there are reasons to assume that they may have fewer problems with alcohol than those who also have other problems covered by official records. The same kind of reasoning may hold true for those registered for criminality and psychiatric care.

In order to understand the process behind maladjustment in general, more finely grained analyses, using detailed information available in the project about each of the aspects of maladjustment, are needed and will be performed. For some of the issues illuminated by data from official records, the presentation was restricted to data for males. Further analyses will, of course, also include data for the females.

Chapter 6

BIOLOGICAL MATURATION AND LIFE STYLE AMONG FEMALES: A SHORT-TERM AND A LONG-TERM LONGITUDINAL PERSPECTIVE

INTRODUCTION

A theme in this study is that a person's life situation cannot be adequately understood without mapping the interplay of biological, psychological, and sociocultural factors in development. In this chapter, some empirical illustrations of this theme are presented with particular reference to the part of the life cycle from middle adolescence to adulthood for females. It is the analysis of the specific interaction of different factors, simultaneously operating on different levels, that enables understanding of why some persons choose their ways of living, not the attention to particular potentially predisposing factors per se, whether organismic or environmental. Particular attention is paid to the response of others to changes in the individual and the way in which characteristics of the person, in conjunction with characteristics of the interpersonal network in adolescence, shape the direction of the individual's future life.

The point of departure for the analysis presented here is the assumption that the rate at which biological maturation takes place is an important and often ignored factor in the study of social developmental processes (Mussen & Jones, 1957, Mussen & Young, 1964; Petersen & Taylor, 1980). It will be argued that these differences in biological maturity among girls in middle adolescence and subsequent interactions with the environment to some extent set the stage for the kind of life the girls live in this period as well as in future life. It is shown that some social behaviors that typically appear among girls in mid-adolescence are closely connected to the level of physical maturation. They occur more frequently and more intensely among those who mature early than among those who mature late. Physical maturation does not exercise a direct effect on the expression of these behaviors but must be seen in the light of the social network of the girls in this period. It is shown that biological maturity is related to social adjustment processes in middle adolescence and that the effect is mediated by characteristics of the girl's circle of peers. It is particularly the subgroup of very early mature girls who have formed close contacts with older peers that is responsible for the observed behavior differences between early and late maturers. A closer analysis of the role of peers as norm transmitters shows that peers' *reactions* rather than their general *evaluations* are an operating factor.

The question of whether the observed behavior differences in middle adolescence are relatively transient or lasting is examined using self-reports and register data extending into adulthood. It is shown that the developmental path for these specific behaviors follows a model for longitudinal consequences in which the impact of physical maturation is general when the girls enter puberty; however, differences in growth rate determine the point in time at which the behavior sequence is activated. Hence, observed behavior differences between early and late developing girls in mid-adolescence are related to the timing of the onset of the behaviors, with the ultimate result that the girls who mature late eventually catch up with the others.

What then about lasting consequences of differential physical maturation? It is argued that the biological maturity stage and the interpersonal relationships found during a limited period in the girls' lives in middle adolescence can have profound effects on and set the direction for the kind of life the girls aspire to in adulthood. Despite the fact that the observed behavior differences for some types of behavior between girls become lost after a short period of time, they leave traces in life style orientation that can be found in the adult life of the girls such as whether the girls choose a traditional family life style or concentrate on education and career.

The empirical part of the chapter has four purposes:

1. Data are presented concerning the influence of biological maturity on some teenage behaviors often assumed to indicate a propensity toward further negative social adjustment. Norm violations in mid-adolescence will be investigated for girls at varying stages of biological maturity.
2. In order to elucidate factors that mediate the influence of biological maturation on norm-breaking behaviors, peer relations will be introduced as a moderating factor.
3. The issue of lasting individual differences is first investigated in the short-term perspective for use of alcohol. With respect to long-term consequences, alcohol abuse and crime at adult age are compared for females with varying levels of biological maturity in adolescence.
4. Given the hypothesis that girls with varying combinations of physical development and interpersonal networks in mid-adolescence are likely to

choose different future life style orientations, marriage rate, number of children, and educational status are compared among females with varying combinations in adult life.

BIOLOGICAL AGE AND NORM VIOLATIONS IN ADOLESCENCE

Criminological research and the literature on social adaptation in adolescence show a marked increase in breaches of norms during the early teenage years (Jessor & Jessor, 1977; Stattin & Magnusson, 1984). The growing number of violations as chronological age increases has been interpreted in part as a process in which the individual establishes an independent norm system. The developmental forces that encourage conformity and conventional rules gradually lose their influence after the early years of youth. In dealing with the issue of variations in norm-breaking behaviors among individuals in this period, the conventional procedure in empirical research has been to view norm-breaking as an individual difference variable and relate it to other information about the individuals under study. Correlates of norm-breaking have then been sought among ecological factors, such as upbringing conditions, social networks, school adjustments (Brooks-Gunn & Petersen, 1983; Jessor, 1984; Jessor & Jessor, 1977). However, when comparing subjects of the same chronological age and relating these individual differences to concurrent and later data, it is not certain if, during the particular occasion in question, the same psychological processes are being measured across individuals. Research on physical growth has made it clear that the assumption of homogeneous development does not always hold true; hence, chronological age cannot be used as the only meaningful reference scale for development (Goldstein, 1979; Magnusson, 1985b; Peskin, 1967).

The starting point for the empirical data presented here was a study testing the hypothesis that certain social behaviors typically originating in adolescence are related to biological maturation levels (Magnusson & Stattin, 1982). This implies that individual differences in adjustment among girls at a particular point in the adolescent period partly represent biological time-lag effects.

Data for Biological Maturation

The relation between norm-breaking in mid-adolescence and rate of biological maturation was investigated for a group of girls for whom complete menarcheal data were obtained when they were about 15 years of age in the 8th grade in 1970. Five hundred and eighty-eight girls were enrolled in the school system, and data on menarche were obtained for 509 girls who were present at school on the days when data were collected. Most of the 509 girls were born in 1955. However, 10 (1.9%) were born in 1956, and 43 (8.4%) started school late or had

TABLE 6.1.
Means and Standard Deviation on Normbreaking for Four Menarcheal Groups of Girls and Tests of Mean Differences Between the Groups of Girls

	Age at Menarche								Mean Differences	
	−11 (n = 48)		11–12 (n = 98)		12–13 (n = 178)		13− (n = 112)			
Normbreaking	*M*	*sd*	*M*	*sd*	*M*	*sd*	*M*	*sd*	*F**	*p*
Home										
Ignore parents' prohibitions	2.40	1.14	1.93	0.93	1.89	0.93	1.89	0.93	4.84	$<.01$
Stay out late without permission	2.67	1.28	2.08	1.12	1.96	0.92	1.74	0.87	9.96	$<.001$
School										
Cheat at an exam	2.19	1.10	2.05	0.98	2.09	0.94	2.00	0.94	0.48	.70
Play truant	2.77	1.59	2.08	1.19	1.74	1.00	1.74	0.98	12.40	$<.001$
Leisure time										
Smoke hashish	1.13	0.33	1.04	0.20	1.01	0.11	1.01	0.09	6.65	$<.001$
Get drunk	2.65	1.55	2.14	1.39	1.75	1.13	1.54	1.00	11.43	$<.001$
Loiter in town every evening	2.23	1.28	2.05	1.08	2.01	1.09	1.75	1.00	2.69	$<.05$
Pilfer from a shop	2.02	1.18	1.64	0.98	1.59	0.91	1.50	0.82	3.64	$<.05$
Total normbreaking:	2.23	0.90	1.88	0.71	1.76	0.55	1.63	0.55	10.57	$<.001$

*df = 3/432

not advanced in the ordinary manner. In order to avoid any confounding with chronological age, data were analyzed for only those 466 girls in grade 8 who were born in 1955.

Age at menarche was measured by an item in an inventory administered in April 1970 when the average age of the girls was 14:10 years (14 years and 10 months). The median age of self-reported menarche was 12:86 years, which corresponds closely to national figures for that age cohort (Lindgren, 1976).

The girls were grouped into four menarcheal groups: (a) menarche before the age of 11; (b) menarche between the ages of 11 and 12; (c) menarche between the ages of 12 and 13; and (d) menarche after age 13.

Norm Violations in Girls of Varying Biological Maturation

Data on norm violations were collected at the average age of 14:5 years by means of a norm inventory. Subjects' answers concerning their actual breaches of norms were analyzed for frequency of norm violations up to the point in time that the instrument was administered.

The relation between biological maturity as measured by age at menarche and norm violations at the age of 14:5 years is presented in Tables 6.1 and 6.2. Table 6.1 presents means of norm-breaking for the girls subdivided into the four menarcheal groups, together with statistical tests of differences among the four groups of girls. Table 6.2 shows the percentage of girls in each one of the four groups who reported frequent norm-breaking (four times or more) for each situation described.

TABLE 6.2.
Percentage of Girls in Different Menarche Groups Reporting Frequent Normbreaking at 14:5.

	Age of Menarche			
	−11 (n = 43)	*11–12 (n = 98)*	*12–13 (n = 178)*	*13− (n = 112)*
Home				
Ignore parents' prohibitions	16.7	7.1	2.8	3.6
Stay out late without permission	27.1	12.2	5.6	4.5
School				
Cheat on an exam	17.0	5.1	5.1	7.3
Play truant	39.6	14.3	5.6	7.1
Leisure time				
Smoke hashish	12.0	4.1	1.1	0.9
Get drunk	35.4	20.0	7.9	6.3
Loiter in town every evening	20.8	9.1	8.5	3.6
Pilfer from a shop	14.6	5.2	4.5	1.8

As can be seen in Table 6.1, there is a clear association between age of biological maturation and reported frequency of norm-breaking, meaning that norm violations at the time of the testing were related to how far the girls had come in their physical maturation. As can be seen in Table 6.2, there was a considerably higher percentage of early maturing girls who reported frequent violations of norms. It is also obvious that the relation is not linear; the earliest-developing girls differed markedly in mean and frequency of norm breaches from the groups of later developing girls.

Older Peers as Social Mediators

The higher norm-breaking frequency among early developing girls than among the late developing raises questions as to the type of factors that contribute to the systematic relation between biological maturation and the tendency toward norm-violations. It is unlikely that a single direct causal link exists between physical or hormonal changes and social behaviors. On the contrary, we believe that this effect is most likely mediated by an environment that changes often and to a large extent as a consequence of changes in the individual (Magnusson & Allen, 1983b). This proposition requires some form of analysis of the effects of physical maturity.

It is reasonable to assume that a biologically early matured girl will be considered by people in her surroundings as older than she actually is. The expectations and demands upon her may be different from those placed on her late-matured peers. One can also expect the early-matured girl to associate more with chronologically older persons, signifying new and more advanced habits and leisure-time activities. In her association with older peers, the girl may encounter the more tolerant attitudes towards norm-breaking that characterize older groups of teenagers. Through association with them she will more often be likely to meet and participate in situations that may lead to rule-breaking and encounter more positive attitudes towards norm violations.

For the present groups of girls, Magnusson, Stattin, and Allen (1986a,b) showed that the early-matured girls sought out and were sought out by others who were congruent with their early biological stage of maturity. Seventy-four percent of the earliest matured girls reported having older friends whereas only 39% of the latest developed girls said that they had older friends at age 14:5 years. Results from sociometric ratings also showed that the earliest developed girls were significantly more often apt to nominate other early-matured girls as friends, whereas the late-matured girls more often nominated girls who, like themselves, were late developers. In addition, the early-matured girls were found to be considerably more advanced in their contacts with the opposite sex. Eighty-three percent of the early-matured girls had been or were going steady with boys as compared to 52% of the late-matured girls. Moreover, more than four times

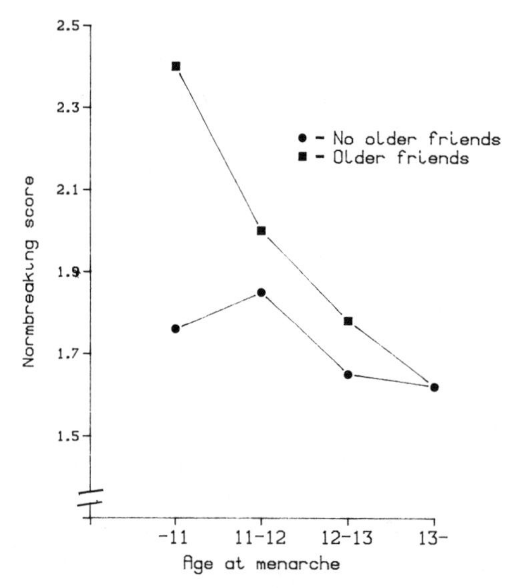

FIGURE 6.1. Norm-breaking scores for girls in four menarcheal groups. The girls are separated into those with and those without older friends.

(45%) as many girls among the earlier developed group had had sexual intercourse with boys at age 14:5 years as compared to the later matured.

The finding that early-matured girls were more oriented towards older age groups of peers (and towards girls of the same chronological age who matched their early development), suggests the hypothesis that the relation between early maturation and high norm-breaking is mediated by the association with more mature, older peers. The relation between a tendency toward norm violation and menarcheal age for those 14:5 year old girls with and those without older friends is shown in Figure 6.1.

The statistical analysis of the results shown in Figure 6.1 revealed a significant mean difference between girls with and girls without older friends ($p<.01$). For the girls who reported having no older friends at this age, there was no significant difference in norm-breaking among the menarcheal groups (F 3,176 = 1.31, n.s.). For girls who reported having older friends, a clear and significant difference in norm violations among the menarcheal groups was found (F 3.183 = 9.14, $p<.001$). In accordance with the hypothesis, the difference in means between girls with and those without older friends is mainly explained by the difference between very early matured girls with and very early matured girls without older friends ($p=.059$). This elucidates the results reported in Tables 6.1 and 6.2 that showed that the group of very early developed girls differed from the other groups in both frequency and intensity of norm violation. For the other menarcheal groups, the difference was in the expected direction—having older friends was associated with more norm-breaking—but the differences were small and insignificant. The means for norm-breaking were almost the same for girls with and for girls without older friends in the group of late developed girls.

Peers as Norm Transmitters

Up to this point, the results may be summarized as follows. For girls at 14:5 years of age, there are marked differences in norm violations that are related to the girls' social network, particularly the association with more mature, older friends. The main effect at age 14:5 of having older friends is concentrated in the group of very early developed girls. At this point, the role that friends play as norm transmitters (based on how the girls perceive their friends' opinion of norm-breaking and the reaction they expect after their own norm violations), comes into focus.

In order to investigate the role of peers as norm transmitters, data were used on peer evaluations and expected peer sanctions from a norm inventory administered at age 14:5 years (see Magnusson et al. 1975, pp. 95–120). It was stated in

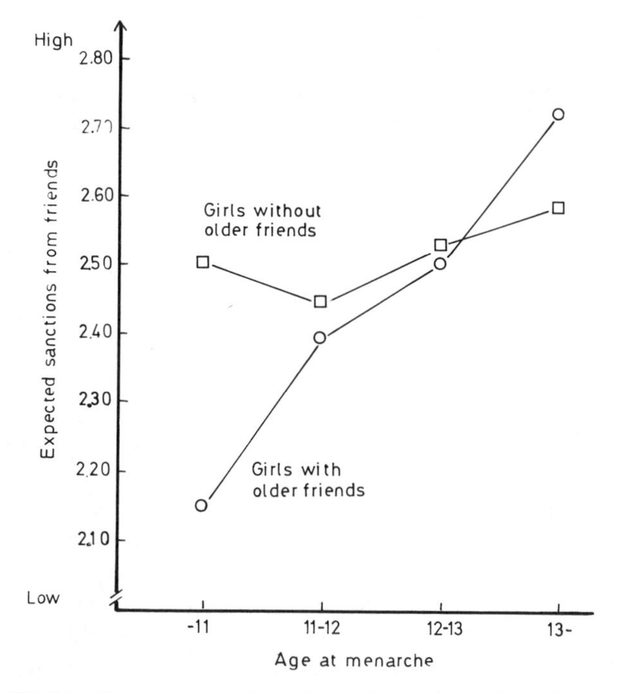

FIGURE 6.2. Expected sanctions from friends for girls in four menarcheal groups. The girls are separated into those with and those without older friends.

the instrument that: "By 'peers' we mean those whose opinion you care most about, whether they are in your class, in a 'club', or your best friend."

For the girls' judgments of their peers' evaluations of normbreaking, there were no differences between the four menarcheal groups of girls ($F_{3,395} = 1.11$, n.s.). This result suggests that girls generally are confronted by similar evaluation systems among their peers. On the other hand, the results on expected peer sanctions show the same pattern found for the relation between biological maturity and norm violation frequency. Biologically early-matured girls expected weaker reactions after norm violations than did late-matured girls ($F_{3,370} = 3.18$, $p<.01$). When the age of friends was introduced as a moderating variable, the relation between biological maturity and expected peer sanctions appeared in the expected direction significantly only for those girls with older peers ($F_{3,138} = 4.65$, $p<.004$). The results are graphically depicted in Figure 6.2. It is interesting to note that the latest matured girls with older friends tend to expect stronger reactions than late-matured girls without older friends.

Interpretation: Deviance or Differential Growth Rate?

The results presented raise an important problem with respect to the interpretation of the role of girls' early physical maturation in their social adjustment (Eichorn, 1975; Livson & Peskin, 1980). At first glance, they seem to indicate that very early-maturing girls form a group that runs a higher risk of later maladaptation than later maturing girls. However, the results can be interpreted in two different directions, as follows (Magnusson & Stattin, 1982; see also Chapter 3).

1. Frequent norm-breaking among early-matured girls is an indication of a more general pattern of personal maladaptation, directly or indirectly related to early physical maturation. What is manifested in behavior is then an underlying more "deviant" orientation on the part of these early-developed girls.

If this interpretation of the influence of biological maturation on later social development is valid, it can be predicted that different indicators of social maladaptation in late adolescence and in adulthood will show a significant relation to the age of the menarche.

2. Frequent norm-breaking at the age of 14 among early-matured girls is only an expression of the fact that these girls enter the normal process of norm-transition earlier than late-matured girls. This interpretation implies that all girls pass through the same process; the difference is that early-matured girls pass through this process at an earlier age than do their late-matured peers. It suggests that individual differences among girls are due mainly to the timing of the onset of the behaviors in question. Hence, differences in growth rate would not be expected to give persistent or long-term consequences.

If this hypothesis is valid, the late-matured girls will catch up with the early-developed. The relation between norm violation in later life stages and the age of menarche will attenuate and in an extreme case approach zero. Individual differences in norm violations at late adolescence and adulthood are then determined by person-bound factors other than the onset of menarche and/or environmental factors unrelated to physical maturity during puberty.

BIOLOGICAL MATURATION AND DEVIANCE IN A LONGITUDINAL PERSPECTIVE

Short-term Analyses

In order to investigate this problem for one particular type of normbreaking, data on the use of alcohol were analyzed at two age points in adolescence, separated by 17 months. In Table 6.3, data on alcohol use at the average age of 14:5 years are shown, and data collected at 15:10 years are presented in Table 6.4.

Tables 6.3 and 6.4 represent a temporal sequence and as such contain interesting information about trends over time and inter-group differences in the total group of girls. As expected, more girls reported at the latter test occasion that they had been drunk. At age 14:5 years, about 40% of the girls reported having been drunk. At age 15:10 years, the same figure was 70%. The net increase for the categories of rather excessive drinking was about the same for early- and late-matured girls from the age of 14:5 years to 15:10 years. For example, in the earliest developed group the percentage that had been drunk more than 10 times

TABLE 6.3.
Percentage of Girls in Four Menarcheal Groups with Varying Frequency of Drunkenness up to Age 14:5.

	Frequency of Drunkenness					
Age at Menarche	*Never*	*Once*	*2–3 Times*	*4–10 Times*	*> 10 Times*	*N*
-11	37.5	12.5	14.6	18.8	16.7	48
11–12	51.0	13.3	15.3	11.2	9.2	98
12–13	62.1	13.0	16.9	3.4	4.5	177
13–	71.4	10.7	11.6	4.5	1.8	122
N	258	54	65	31	27	435
%	59.3	12.4	14.9	7.1	6.2	

$Chi^2 = 39.88$ df $= 12$ $p < .001$

TABLE 6.4.
Percentage of Girls in Four Menarcheal Groups with Varying Frequency of Drunkenness up to Age 15:10.

	Frequency of Drunkenness					
Age at Menarche	*Never*	*Once*	*2–3 Times*	*4–10 Times*	*> 10 Times*	*N*
−11	25.0	7.5	5.0	22.5	40.0	40
11–12	25.7	5.9	18.8	20.8	28.7	101
12–13	29.4	10.0	19.4	18.8	22.4	170
13–	35.3	9.2	20.2	11.8	23.5	119
N	128	37	78	76	111	430
%	29.8	8.6	18.1	17.7	25.8	

$Chi^2 = 15.71$ $df = 12$ $p = .20$

increased from 16.7% to 40% over the 17-month period. For the latest-matured group of girls, the percentage increased from 1.8 to 23. Despite this equality in net increase of excessive drinking, a tendency can be discerned for the late developing girls to catch up with the early ones with respect to drinking per se. The proportion of girls that moved from never having been drunk to having been drunk at least once is considerably higher for the late-matured girls (from 28.6% to 64.7%) than for earlier developers (from 62.5% to 75%). Thus, in terms of moderate drinking the "spurt" for the late matured girls comes during this 17-month period.

In further analyses, a complementary two-way ANOVA was computed with the test occasion as a trial factor and menarcheal age as a grouping factor. The dependent variable (drunkenness) was dichotomized; subjects who had not been drunk scored zero and those who had been drunk at least once scored one. There was a significant main effect of menarcheal age ($F = 3.44$, $p<.05$) and of test occasion ($F = 104.97$, $p<.001$). In addition there was a significant interaction between the test occasion and menarcheal age ($F = 2.59$, $p<.05$), reflecting the increase in norm-breaking among late-matured girls.

To summarize the obtained results, Figure 6.3 depicts growth curves with respect to having been drunk at least once for the four menarcheal groups of girls over the adolescent period. In this figure we have also included data on drunkenness for the age point 14:10 years.

For the earliest matured group of girls, the maximum frequency for drunkenness is at 14:10 years, since the proportions of girls who had been drunk at 14:10 years were also the same at 15:10 years. The figure also reveals clearly the catch-up for the other groups of girls. This is particularly evident for the latest-matured group of girls.

Long-term Analyses

The figures for the development process during adolescence presented in the preceding section indicate that the influence of biological maturation on alcohol habits seems to be restricted to a rather short period in adolescence, which supports a timing interpretation of the relation between biological maturation and conduct among females during adolescence. In order to determine if the observed differences with respect to drunkenness between the menarcheal groups of girls would be reflected in alcohol habits at an adult age, self-reported alcohol habits and registered alcohol abuse were gathered when the girls were around 26 years of age.

In Table 6.5, the self-reported data at adult age for the frequency of alcohol consumption are related to the age of menarche. No significant relation was obtained when frequency of alcohol consumption was used as the dependent measure (this is also the case for amount of alcohol consumed, see Magnusson, Stattin, & Allen, 1986a). Females who report that they drink most frequently at adult age come as often from the group of very early-matured girls as from the group of very late-matured girls. These results confirm the tendency suggested by those data obtained at age 15:10, namely, that differences in alcohol use among the menarcheal groups of girls occur for a limited period of time.

The same conclusion can be drawn when inspecting alcohol abuse as measured by register data. Only 15 girls, (3.3%) of the 458 girls with complete register data on alcohol abuse, were registered up to age 26. Of the four groups,

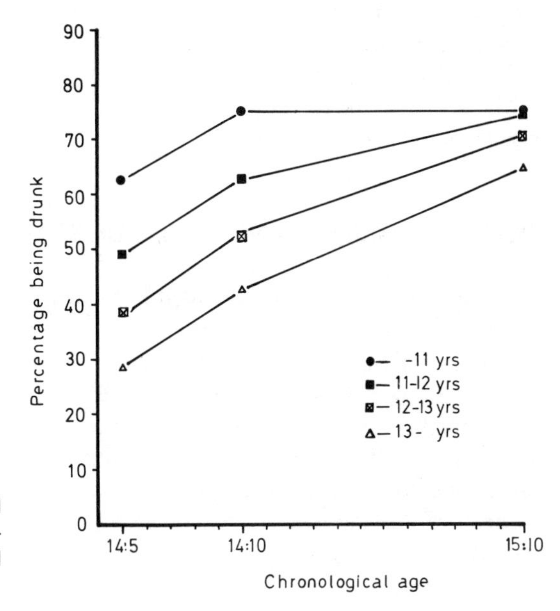

FIGURE 6.3. Developmental trends with respect to intoxication for girls in four menarcheal groups.

TABLE 6.5.
Percentage of Girls in Four Menarcheal Groups with Varying Frequency of Alcohol Consumption at Age 26.

Age at Menarche	*Never*	*At Least Sometimes*	*Weekly*	*N*
	Alcohol Consumption			
-11	5.7	82.9	11.4	35
11–12	7.1	79.8	13.1	84
12–13	9.0	75.2	15.9	145
13–	11.2	77.5	11.2	89
N	31	274	48	353
%	8.8	77.6	13.6	

$Chi^2 = 2.58$ df = 6 p = .86

4.3% of the girls with menarche before age 11 were registered by the police, social welfare authorities, or by psychiatrists for alcohol problems; 5.6% of the girls who had menarche between age 11 and 12; 0.5% of the girls with menarche between 12 and 13; and 5.0% of the latest developed group of girls. There were no significant differences between the four menarcheal groups of girls.

A broader analysis of the issue of the persistence of norm-breaking was performed by Magnusson, Stattin, and Allen (1986a,b). They compared early- and late-matured girls with respect to registered crime up through age 26. There was a tendency for the early-matured group of girls to be over-represented among the girls with criminal records. Nevertheless, in general, the results failed to provide strong evidence for the argument that menarcheal age is connected with more persistent antisocial behavior.

BIOLOGICAL MATURATION AND FUTURE LIFESTYLE ORIENTATIONS

The results that have been presented in this chapter on the relation between the rate of biological maturation and indicators of social maladjustment in terms of excessive drinking and criminality indicate that early biological maturation does not have long lasting consequences in those respects. However, it would be premature to conclude that the differences in social behavior between the menarcheal groups that were observed in mid-adolescence do not represent a distinct underlying psychological process at this stage of development that might be carried over to later periods. This is to say that individual differences in norm-breaking behavior might have another "growth function" (McCall, 1977) than reflecting a deviance motive.

The lack of systematic long-term consequences of early biological maturation with respect to antisocial behavior in general suggests a different interpretation of the findings. This interpretation involves taking into account both the early-matured girls' adaptation to adult roles and responsibilities and their accommodation to the prevailing norm climate among preferred peers. Rather than viewing the higher frequency of norm-breaking among early-developing girls as a risk factor for later social maladaptation, behaviors such as ignoring parents' prohibitions, staying out late without permission, alcohol drinking, and loitering in town in the evenings, might be interpreted as a movement away from dependence on the home environment and toward the establishment of an identity as an independent adult. The exposure to older teenagers' more tolerant reactions to norm-breaking (especially older boys and boyfriends) constitutes another factor that further contributes to the early-maturing girls' greater likelihood of breaking conventional rules during this time period. This behavior, interpreted negatively and with a deviant connotation by the adult world may be, from the perspective of the growing girl, a logical step in her development toward an adult.

An interpretation of the more frequent violations of norms in mid-adolescence among early matured girl in terms of an earlier acquisition of an adult social role leads to a number of testable hypotheses concerning the future of these girls.

Self-Perception, Family Life, and Own Children

The girls' perception of themselves is a crucial factor for the analysis of the long lasting effects of early maturation. A reasonable hypothesis is that an early-matured girl in this period views herself as psychologically more mature than her later-maturing peers. Preliminary findings point in the direction of this acquisition-of-adult-social-role hypothesis. At the age of 14:5 years, the girls were asked if they considered themselves more or less mature than their classmates; 42% of the earliest biologically matured girls perceived themselves as more mature and only 2% as less mature. The situation was quite different among the late-matured girls. Only 16% of the latest developed girls considered themselves more mature than their classmates, and nearly 20% thought of themselves as less mature. The relationship between age at menarche and self-perceived maturity was significant at a high level. The results are presented in Table 6.6. This finding is in agreement with results presented by others. Faust (1960), for example, reported that early matured girls generally scored higher on the items "has older friends" and "seems grown-up" (see also Simmons, Blyth, & McKinney, 1983).

In response to a question at age 14:5, more than half of the early-developed girls said that they felt themselves to be different than their peers, whereas a minority, 17%, among the latest developed girls answered in the affirmative ($chi2 = 52.8$, $df = 6$, $p<.001$). On a complementary question, nearly half of the early-developed girls who considered themselves as different stated that this was so because they felt themselves more romantic than other girls.

That early biologically matured girls act and psychologically feel themselves

TABLE 6.6.
Self-Perceived Maturity Among Girls at Age 14:5.
(Percentages.)

	Do You Feel More or Less Mature Than Your Classmates?			
Age at Menarche	*More Mature*	*About as Mature*	*Less Mature*	*N*
-11	41.7	53.6	2.1	48
11–12	37.1	60.8	2.1	97
12–13	20.9	75.7	3.4	177
13–	16.1	64.3	19.6	112

$Chi^2 = 52.8$ $df = 6$ $p < .001$

more grown-up than late developers, does not say anything about the kind of grown-up life to which these girls aspire. This important question of whether differential maturation in adolescence carries implications for future life styles will now be discussed.

Earlier results showed that early maturers associated more with older peers, had more stable relations with boys, and were more sexually active. It could consequently be hypothesized that the future life for the early developed girls primarily means establishing a traditional family life. If this is the case, we would expect that girls who mature early would be more likely to look forward to starting their own family. We would also expect a higher proportion of these girls to have stable relations and bear children at an early age. Since the impact of peers was found to be particularly strong for the very early-matured girls, it could be assumed that starting a family would be of particular concern for this subgroup of girls.

Marriage and Child-bearing. In order to determine if there were differences between early and late-matured girls with respect to attitudes towards child-bearing, at 14:10 girls were asked whether they looked forward to bearing and bringing up children. More than two out of three girls among the earliest developed wished this very much or relatively much in comparison with 42% of the latest developed girls ($chi^2 = 22.48$, $df = 6$, $p<.05$).

In connection with a follow-up assessment at adult age, when the girls were, on the average 26 years old, they were asked if they were married or lived in a stable relationship with a man. About as many early developed females as late developed lived in a stable relationship (see Backenroth & Magnusson, 1983). Data on living-together status in this case referred to the particular age point of 26 years and did not cover all the different family-type relations that might have occurred at earlier points in time. However, comparing the number of children the four menarcheal groups of females had at this point in time, a considerably higher percentage of the early developed females had given birth to one or several children than the late developed ones ($chi^2 = 14.11$, $df = 3$, $p<.01$).

Support for the hypothesis that biological maturation interacts with the social network in mid-adolescence to produce future life effects was obtained when the females were divided into those with and those without older peers in mid-adolescence. As depicted in Fig. 6.4, the number of children for the most early developed group of females was higher for females with older peers in mid-adolescence than for the early developed females without older friends. No impact of older friends at mid-adolescence could be traced for the other groups of females.

Education

Indications from the educational-vocational domain further support the conclusion that the kind of grown-up life to which the early developed girls aspired in adolescence was a traditional family life style rather than a nontraditional vocational career.

Two observations from the data gathered when the subjects were in the compulsory school suggested that long-term differences between the menarcheal groups of girls would occur for educational aspirations. The indications were that early developers would drop out from school early and the late developing girls would be more likely to go on to an educational career.

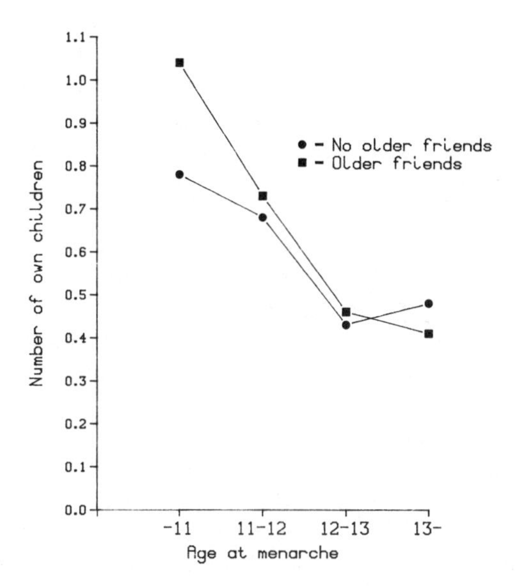

FIGURE 6.4. Number of children at age 25:10 for women grouped according to menarcheal age and contacts with older friends at mid-adolescence.

First, consider school discipline problems at the age of 14:5 years. As shown in Tables 6.1 and 6.2, early-matured girls violated school regulations more often than did late-matured girls. Almost 40% of the earliest developed girls had played truant often at school during this period, in comparison with the 7.1% of the latest-matured girls who reported having done this often. Second, greater school maladjustment was found among early-matured girls in an earlier study in the project. Andersson, Dunér, and Magnusson (1980), using data from the project, compared a group of early-matured girls with a normative control group of the same chronological age on various aspects of school adjustment at the ages 13 and 15. These included feelings of dissatisfaction at school, educational aspirations, burden of work, as well as ratings of concentration difficulties, school motivation, and demands made by teachers. At both ages, early-matured girls showed more negative attitudes towards school and school work, and they were rated more unfavorably by their teachers. The differences in school adjustment between the two groups of girls were sharper at the latter age. These two observations, together with a higher frequency of school problems and lower achievement among early-matured girls as reported in earlier literature, suggest long-term consequences for the further education of the girls (Davies, 1977; Frisk, Tenhunen, Widholm, & Hortling, 1966; Simmons, Blyth, & McKinney, 1983; Simmons, Blyth, van Cleave, & Bush, 1979). Therefore, a comparison was made of the educational status in adult life among the menarcheal groups of females.

The educational level of the menarcheal groups was compared using data from the mail inventory administered at the average age of 26 years. The girls' answers to a question on their current level of education were coded into four categories: (a) Compulsory school education, (b) Higher secondary school education; practical, (c) Higher secondary school education; theoretical, and (d) Academic and college education. Table 6.7 presents the educational status of the females at age 26, grouped according to menarcheal age in adolescence.

As can be seen in Table 6.7 there were marked differences among the four menarcheal groups of females in level of education at adult age. The differences were highly significant ($p<.01$). Most striking was that a minority of the most early-developed girls (27.9%) had some form of theoretical education above the obligatory nine-year compulsory schooling, whereas a majority of the latest-developed females (60.0%) had such an education. Only 2.3% of the earliest-matured girls had entered a college or university in comparison with 12% to 15% among the other groups of females. Magnusson, Stattin, and Allen (1986a) have presented the results of control analyses that investigated the unique contribution of menarcheal age to the variation in adult educational status over and above other standard predictors of educational achievement, that is, the educational status of the parents and the girls' intelligence. A stepwise multiple regression analysis, with parents' education, the girls intelligence level, and their menarcheal age as independent variables, yielded a multiple correlation of .49

TABLE 6.7.
Level of Education at Adult Age for Girls in Four Menarcheal Groups.

Age at Menarche	*Academic or College Education*	*Higher Secondary School, Theoretical*	*Higher Secondary School, Practical*	*Compulsory Education*
−11	1	11	13	18
	(2.3%)	(25.6%)	(30.2%)	(41.9%)
11–12	13	39	20	28
	(13.0%)	(39.0%)	(20.0%)	(28.0%)
12–13	21	86	25	34
	(12.7%)	(51.8%)	(15.1%)	(20.5%)
13–	16	47	19	23
	(15.2%)	(44.8%)	(18.1%)	(21.9%)
N	51	183	77	103
%	12.3	44.2	18.6	24.9

$Chi^2 = 21.83$, $df = 9$, $p = .009$

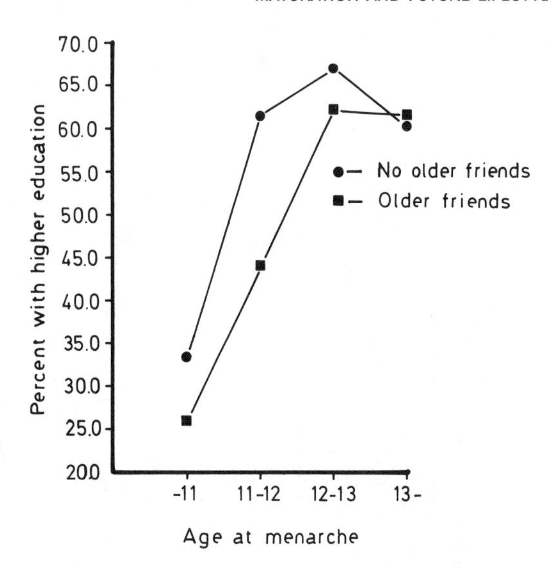

FIGURE 6.5. Percentage of women with higher education grouped according to contacts with older friends at midadolescence.

($p<.001$). The three independent variables contributed significantly to the variation in adult education. A partial correlation coefficient of .15 ($p<.01$) between menarcheal age and adult education (when controlling for intelligence and the educational status of the home) indicated that most of the variance of menarcheal age in common with adult education was unique (the raw correlation was .17).

A further investigation addressed the role of peers in mid-adolescence, subdividing the girls in the four menarcheal groups into those with and those without older peers in mid-adolescence. The percentage of girls in the eight subgroups with higher theoretical education is depicted in Fig. 6.5.

A very similar picture to that presented earlier for older friends is obtained for educational status in adult age. To have had older peers in mid-adolescence has a profound impact for the early developed girls. A considerably lower percentage of the two earliest developed groups of girls with older friends at 14:5 years continued with a higher theoretical education than did early developed girls without older peers (F 1,123 = 4.20, $p<.05$). As expected, little impact of older friends was established for the late matured girls. Figure 6.5 shows very large differences in adult educational status among the groups of girls. More than two times as many of the late-developed girls went through a higher theoretical education as did early-developed girls who associated with older peers in mid-adolescence.

COMMENTS AND CONCLUSIONS

In this chapter studies of four issues have been reported: (a) the role of biological maturation in norm-breaking behavior among girls during adolescence, (b) peer relations as a social mediator and modifier in the socialization process, (c) the short- and long-term consequences for norm-governed behavior of early biological maturation, and finally (d) the impact of the interaction of biological maturation and the girls' social network on future life styles. These issues are dealt with in further, more finely graded analyses by Stattin and Magnusson (in press a,b).

At the time of the test occasion in mid-adolescence at the age of 14:5 years, early-maturing girls were found to have violated various types of norms considerably more frequently than their late-matured peers. A hypothesis was advanced that biological maturity is related to social adjustment by way of the girls' own choice of peers. In accordance with the hypothesis, it was found that differences in norm violations occurred for those girls who had formed close contacts with older and more mature peers. The impact of having older friends was concentrated in the group of very early matured girls. Illuminating the role of friends as norm transmitters, the girls' conceptions of their peers' sanctions in relation to their own norm violations was found to have the same pattern as the relation between biological maturity and norm violation frequency.

The third problem investigated was the question of the longitudinal implications of the impact of menarche. In the short-term perspective, results for one type of behavior (alcohol use) showed that the strong relation found at age 14:5 years between biological maturity and frequency of drunkenness was considerably decreased by the age of 15:10 years. At an adult age no systematic correlation between alcohol consumption (self-reported alcohol use or registered alcohol abuse) and the age of menarche existed.

A hypothesis was advanced suggesting that differences in social experiences in the adolescent period, particularly those connected with the opposite sex, might give rise to different future life orientations among the girls. Available evidence showed that biological maturation and the interactive effect of biological and social influences during mid-adolescence had profound consequences for the life styles of the girls at adult age. The early matured girls were more traditional in their future aspirations, and primarily aimed at establishing a family. This conclusion is suggested by the findings that these girls did not become as engaged in school work as the others, and they left school early. In mid-adolescence they associated considerably more with older peers than did the late developers, had more stable contacts with boys, and were more sexually advanced. They regarded themselves at this time as more mature and romantic than the other girls and looked forward to bringing up their own children. At an adult age, a far higher percentage of the early matured girls had given birth to children than had the late developed girls. By contrast, late-developers more frequently chose a future vocational career. They were more achievement-oriented at school and were more apt to continue with further theoretical education.

The results indicate that the social influences encountered by the girls in the adolescent years (particularly peer reaction to their maturational status) is a factor that affects the future life of the girls. Associating with older friends in adolescence was related both to having children at an adult age and to adult educational status. In both cases, older peer influences played a far greater role for the early-matured girls than for the late developers. Only one out of four in the subgroup of early-matured girls who had close contacts with older peers in mid-adolescence started higher theoretical education above the obligatory nine years in compulsory school, while six out of ten of the late-developed girls did so. Two out of three in the early matured group had children by the age of 26 as compared to one out of three among the late-matured. A connection between educational achievement and menarcheal age and dating was also found by Simmons, Blyth, van Cleave, and Bush (1979), who reported lower grade point averages and lower scores on standardized achievement tests among girls who had begun menstruating, as well as lower achievement results for girls who dated most. One interpretation of these findings is that the early-matured girls' more steady contacts with older boys in the adolescent years reduced the amount of time for studying and other pursuits associated with more advanced education (cf. Simmons, Blyth, & McKinney, 1983). An alternative explanation is that the older peer group with which the early-matured girls associated in these years held lower educational aspirations than did the peers of the late matured girls.

To summarize, the data presented in this chapter indicate that much of the individual difference in norm-governed behaviors that can be observed during female puberty is basically related to individual differences in physical maturity. The data also indicate that these behavior differences were not dependent on the fact of being early matured per se, but that they should be seen as a function of relations to older peers. This finding is an illustration of a central aspect of an interactional perspective on social development. The long-term impact on social behavior is a function of the simultaneous interplay of psychological, biological, and social network factors in middle adolescence; the girls are *both* instigators and products of this process.

Chapter 7

AGGRESSIVENESS, HYPERACTIVITY, AND AUTONOMIC ACTIVITY/REACTIVITY IN THE DEVELOPMENT OF SOCIAL MALADJUSTMENT

INTRODUCTION

One of the central issues of the longitudinal project concerns adult social adjustment. One of the subprojects is devoted to the development of adult criminality, and the investigations reported in this chapter address that problem. They have two theoretical points of departure.

First, most theories on the genesis of adult criminality assume that early misconduct, particularly early aggressiveness, plays an important role. This assumption has been supported by much empirical research (c.f. Loeber & Dishion, 1983; Stattin & Magnusson, 1984). Second, in an interactional perspective the adjustment process cannot be understood and explained without the inclusion of biological factors. The importance of this has been underlined by the results of psychobiological research, which has demonstrated a significant, sometimes strong, negative correlation between extrinsic maladjustment, on the one hand, and low physiological autonomic activity/reactivity as it is reflected in adrenaline excretion, on the other. An interesting problem presents itself here: do those who commit crime as adults differ at an early age from others with respect to their basic physiological functioning? Against this background, the first part of this chapter is concerned with the relationship between conduct and physiological reactions at an early age (about 13 years old) and adult criminality in a longitudinal perspective.

The results of the longitudinal analyses form the basis for the investigations presented in the second part of the chapter. Both ratings of conduct, in terms of aggressiveness and motor restlessness (as an important aspect of the hyperactivity syndrome) and measures of adrenaline excretion at an early age show significant relations to measures of adult criminality. Therefore, it is crucial to identify the relationship among cognitive-affective factors, physiological factors, and conduct in a cross-sectional perspective at an early age, in order to better understand and explain how these factors operate in the developmental process resulting in adult criminality. The second empirical part of this chapter reports cross-sectional analyses of data that were performed in order to elucidate this problem. The study reported here is confined to results for males. Data for females are under exploration and will be reported in a forthcoming volume.

PROCEDURES AND TYPES OF DATA

The analyses, the results of which are presented and discussed here, are built on data for conduct, autonomic activity/reactivity, and a mental task.

Conduct Variables

The teachers rated all their pupils with respect to seven variables, when the boys and girls were around 10 and 13 years old, respectively. Three of these variables were aggressiveness, restlessness, and lack of concentration (The procedure for the data collection was described by Magnusson et al., 1975. The psychometric properties of the teacher's ratings were reported by Backteman & Magnusson, 1981). The results presented here are based on ratings at 13, the same age to which the biological and mental data refer.

Each variable was explained to the teacher by a heading and by description of the extreme manifestations in both directions of the variable under consideration. For the three variables of interest in the analyses reported in this chapter, the extreme negative manifestations were described in the following way:

Aggressiveness. They are aggressive towards teachers and classmates. They may, for example, be impertinent and impudent, actively obstructive, or incite rebellion. They like disturbing and quarreling with classmates.

Motor restlessness. They find it difficult to sit still during lessons. They fidget uneasily in their seats or wish to move about in the classroom, even during lessons. They may be talkative and noisy.

Lack of concentration. They cannot concentrate on their work, but are occupied with irrelevant things, or sit daydreaming. They may work for a few moments but are soon lost in their thoughts again. They usually give up quickly, even when the work is suited to their level of intelligence.

The main characteristics of the rating procedure can be summarized as follows:

1. The ratings were performed on a seven-point scale for each variable.
2. One variable at a time was rated for all students. The names of the students were randomized across variables.

3. Boys and girls were rated separately.
4. The reference groups were boys and girls in their own classes. The teachers were thus instructed to start by choosing the most extreme boy or girl in both directions and mark them on the scale, and then go on with the next most extreme boy or girl, and finally place those who did not deviate in any direction on point 4 of the scale. Adjustment of the general level should be made with reference to characteristics of the class in relation to the characteristic level of the age group.

The statistical effects of using the class as the main frame of reference for the ratings with adjustment of the general level of the class to what the teacher regarded as a "normal" class should be observed. The most probable effect is that the within-class variance has been more reliably reflected in data than the between-class variance.

Autonomic Activity/Reactivity

Adrenaline is a hormone stored and secreted by the adrenal medulla, the inner part of the two adrenal glands just above the kidney. These glands are activated via the sympathetic nervous system (by nerve impulses from the hypothalamus) alerted by perceived external demands of achievement or threat and play a central role in the activation of physiological processes. They interact with receptors of cells in various organs, increase the heart rate and blood pressure, and prompt the release of extra sugar from the liver. Because of its central role in physiological functioning of the total organism, adrenaline excretion was chosen as a principal indicator of an individual's autonomic reactivity to external conditions.

Data for adrenaline in the urine were collected for a sample of boys and girls when they were 13. The procedure was presented in detail by Johansson, Frankenhauser, and Magnusson (1973). Urine was obtained under two standardized conditions: (a) after an ordinary nonexciting lesson at which the pupils were shown a film about ore mining, and (b) after a test-session, involving demand for achievement. These two situations denote a nonstressful and a stressful situation.

Mental Task

Another factor that was considered related to the issue under consideration, was the individuals' ability to concentrate on a mental task. Data were obtained by a test of simple problem-solving, which was administered during the stressful situation when urine was collected. Each item was simple enough to be solved by anyone at all levels of intelligence. The task was to solve as many items as possible in a given time. This type of task was chosen so as to avoid contamination with intelligence as far as possible. This procedure may also ensure that more intelligent boys and girls did not experience higher success with the task.

Criminal Offenses

Data for criminal offenses were obtained from the national and local registers in all communities where participants had lived for shorter or longer periods as reported in Chapters 4 and 5. As a basis for the study of early conduct as an antecedent of adult criminality, the data presented here cover the age period from 18 to 26.

As reported in Chapter 5, the males have been categorized on the basis of the type of crimes. In the analyses presented here, the total group was divided into those who have been registered for a criminal offence during this age period one or several times and those who have not been registered.

Representativeness of Data

In studies of this kind, the representativeness of the data is a crucial matter. It should therefore be noted that for both the conduct ratings and the criminal records there was practically no dropout from the original, representative groups of participants. There was some attrition for data for adrenaline excretion. Because data were available also for the dropouts with respect to other important variables in childhood, we can estimate the possible effects of the drop out and, if necessary, control for them. The analyses thus completed show no indications that the drop out for adrenaline excretion has had any serious effect on the relationships presented here.

CONDUCT, PHYSIOLOGY, AND MALADJUSTMENT IN A LONGITUDINAL PERSPECTIVE

As a first step in an analysis of aspects of individual functioning in the developmental background of adult criminality process, the relations were investigated between conduct at age 13 (in terms of aggressiveness and motor restlessness) and autonomic activity/reactivity (in terms of adrenaline excretion in various situations) at the age of 13, on the one hand, and adult criminality (in terms of registered offences), on the other.

Early Aggressiveness and Adult Criminality

Various theories about the etiology of criminal behavior (including psychodynamic theories, traditional learning theories, and social learning theories) assume that aggressiveness at an early age is an important element in the process underlying adult criminality. This assumption is supported by results from longi-

TABLE 7.1.
Percentage of Subjects with Different Aggressiveness Ratings at Age 13 Who had Registered Offenses at Adulthood (18–26 Years of Age).

	Aggressiveness Score (Age 13)							
	1 *n* = 49	*2* *n* = 74	*3* *n* = 79	*4* *n* = 160	*5* *n* = 91	*6* *n* = 52	*7* *n* = 33	*Total* *n* = 538
No crime	89.8	83.8	79.7	81.3	76.9	55.8	42.4	76.0
Crime	10.2	16.2	20.3	18.7	23.1	43.2	57.6	24.0

tudinal empirical studies in which the relationship between early aggression and juvenile delinquency or adult criminality have been investigated (Feldhusen, Thurstone, & Benning, 1973; Havighurst, Bowman, Liddle, Matthews, & Pierce, 1962; Kellam, Brown, Rubin, & Ensminger, 1983; Kirkegaard-Sorensen & Mednick, 1977; McCord, 1983, Mitchell & Rose, 1981; Mulligan, Douglas, Hammond, & Tizard, 1963; Roff, & Wirt, 1984; West, 1982; West & Farrington, 1973).

The quality of the data in these studies varies and they differ considerably with respect to such characteristics as the representativeness of the samples, the methods used for data collection, the ages to which data for aggression and criminal acts refer, and the time interval between data for aggression and data for criminal behavior. These differences may explain the rather large variation in, for example, the proportion of individuals characterized by early aggressiveness who later become criminals. However, all the studies referred to demonstrate a strong relation between early aggressiveness and later criminality.

In Table 7.1 the percentages of males who had been registered for crimes during the age interval of 18 to 26 are given for each of the rating score categories of Aggressiveness at age 13. The results show that of 85 (16%) boys rated as extremely aggressive (6 and 7), almost half were registered for criminal offenses at adult age. Almost half of them committed serious crimes. On the other hand, of 123 (23%) boys rated as least aggressive at the age of 13, only about 14% were registered and then only a very small portion for serious crimes. These results of a follow-up of a representative group of males, from age 13 to adult age, strongly support the assumption that early aggression is an important antecedent of later delinquency.

Aggressiveness and Motor Restlessness at an Early Age and Adult Criminality

A possible and reasonable interpretation drawn from findings similar to those presented above is that aggressiveness is a main element of early conduct in the etiology of adult delinquency. However, such an interpretation has to be qualified. Empirical research on conduct problems in a general sense suggests that another broad aspect of conduct, covered by the hyperactivity syndrome must be considered (Satterfield, Hoppe, & Schell, 1982). The hyperactivity syndrome is, according to DSM III, characterized by three main components: motor restlessness, concentration difficulties, and impulsivity.

The conceptualization and definition of hyperactivity has been the issue of much debate (Schachar, Rutter, & Smith, 1981). However, factor analyses have been presented that indicate that aggressiveness and hyperactivity constitute separate factors of conduct. They show somewhat different etiology and different implications for later development and later delinquency (c.f. Brocke, 1984; Loney, Langhorne, & Paternite, 1978; McGee, Williams, & Silva, 1984; Milich, Loney, & Landau, 1982; Offord, Sullivan, Allen, & Abrams, 1979; Stewart, Cummings, Singer, & DeBlois, 1981). These findings suggest that an analysis of the *differential relation* between early aggressiveness and hyperactivity on the one hand and adult registered criminality on the other would be useful. The analyses presented here for hyperactivity are based on the teachers' ratings of motor restlessness as one fundamental aspect of the more general syndrome.

The results of an analysis of the differential relation of aggressiveness and motor restlessness to adult criminality are presented in Figure 7.1. There, the percentages of individuals who have committed crimes at an adult age are reported for four categories: (a) Those who were rated neither extremely aggressive nor extremely restless at the age of 13, (b) those who were rated extremely aggressive but not extremely restless at the age of 13, (c) those who were rated extremely restless but not extremely aggressive at the age of 13, and (d) those who were rated both extremely aggressive and extremely restless at the age of 13.

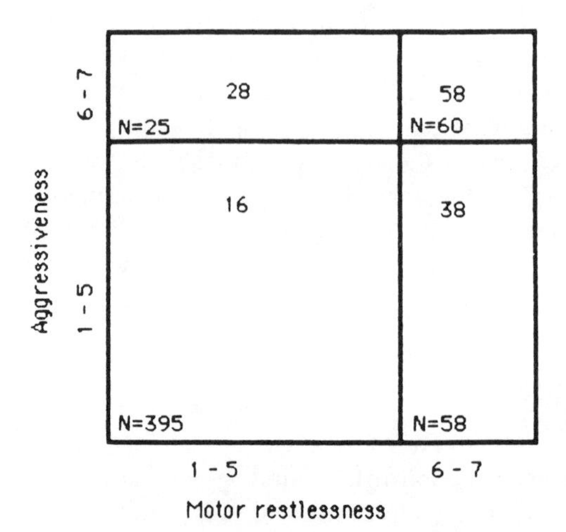

FIGURE 7.1. Percentages of individuals with different combinations of ratings of aggressiveness and motor restlessness at age 13, who have been registered for criminal offenses at age 18–26.

Figure 7.1 demonstrates two tendencies. First, the combination of rated aggressiveness and rated motor restlessness at the age of 13 is a stronger antecedent of adult criminality than the ratings of aggressiveness (see Table 7.1) and also stronger than aggressiveness in the absence of restlessness or restlessness in the absence of aggressiveness. Second, and this is perhaps the most important point, pure motor restlessness seems to be a stronger indicator of later criminality than pure aggressiveness.

Autonomic Activity/Reactivity at the Age of 13 and Adult Criminality

In cross-sectional analyses of data when the subjects were 13 years of age, a negative relation between ratings of aggressiveness and of motor restlessness, on the one hand, and adrenaline, on the other, was found. Thus boys who were rated as high in aggressiveness and in motor restlessness had a lower excretion of adrenaline than other boys. For aggressiveness the coefficients for the correlation with adrenaline excretion were $-.21$ (n.s.) and $-.34$ ($p<.01$) under nonstressful and stressful conditions, respectively. For motor restlessness the coefficients were $-.26$ ($p<.05$) and $-.34$ ($p<.01$) under these conditions.

The negative correlation between conduct variables of this type and adrenaline is compatible with results of much research in psychophysiological laboratory work during the last decade. In general, findings point to a positive correlation between good social and personal adjustment and high adrenaline excretion. This has been shown for relative achievement (Bergman & Magnusson, 1979), for emotional balance and concentration ability in school work (Johansson, Frankenhaeuser, & Magnusson, 1973), for ego strength (Roessler, Burch, & Mefferd, 1967), and for well-adjusted social behavior (Lambert, Johansson, Frankenhaeuser, & Klackenberg-Larsson, 1969). The relationship between delinquency and catecholamine excretion was investigated by Lidberg, Levander, Schalling, and Lidberg (1978). A group of men who had been arrested and were high in psychopathy (as indicated by measures of low socialization, high impulsiveness, and low empathy), showed a conspicuously lower increase in adrenaline and noradrenaline excretion than arrested men who were low in psychopathy, when they were faced with a strong relevant stressor (namely the situation immediately before a trial) as compared to nonstressful situations. For a subgroup of maximum security patients who were convicted for physical violence offenses, Woodman, Hinton, and O'Neill (1977) observed a lower level of adrenaline than for subgroups of mixed offenders and normal control subjects. Olweus (1985) presented data that show a strong significant negative correlation between adrenaline level and a ratings composite of unprovoked aggressive, destructive behavior. Empirical studies by Boydstun and coworkers (1968), Satterfield and Dawson (1971), Satterfield, Cantwell, and Satterfield, 1974, and Klove and Hole (1979) indicated a systematic relation between physiological arousal and hyperactivity which implies that hyperactive children showed signs of low CNS arousal. Taken together, the empirical studies in this area generally demonstrate lower adrenaline levels for subjects with social adjustment problems.

The studies referred to in the preceding paragraph were concerned with the relationship between catecholamines and maladjustment in a cross-sectional perspective. For the understanding of the developmental process antecedent to adult delinquency it is, of course, of interest to investigate the extent to which those who have been registered for crime as adults differ from those who have not been registered with respect to physiological functioning at an early age. In Figure 7.2 adrenaline excretion in the urine, expressed in nanogram per minute, is presented for those who were registered for criminal offence(s) and for those who were not during the age period 18 to 26.

As can be seen in Figure 7.2, the differences in adrenaline excretion between registered and nonregistered individuals are substantial. This result should be evaluated in the light of the rather long interval between the age for the measuring of adrenaline level and the age range to which registered offences refer. This interval also covers a period in the course of individual development characterized by large changes. Considering these circumstances, the differences in

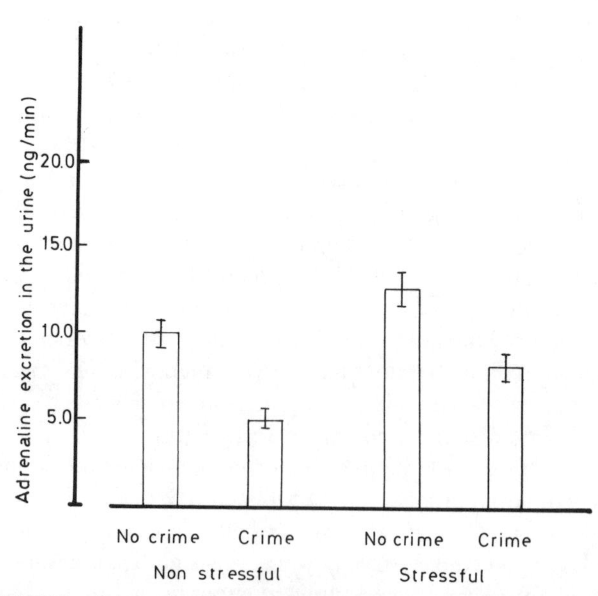

FIGURE 7.2. Adrenaline excretion (nanogram/minute) at age 13 for those who have been registered for crime at age 18–26 and those who have not been registered.

adrenaline excretion at age 13, between those adults who were registered for any type of crime and those who were not registered are surprisingly high.

CROSS-SECTIONAL ANALYSES OF AGGRESSIVENESS, HYPERACTIVITY, AND AUTONOMIC ACTIVITY/REACTIVITY AT AN EARLY AGE

In the longitudinal analyses of data that were reported above, conduct (in terms of aggressiveness and motor restlessness) and autonomic activity/reactivity (in terms of adrenaline excretion) were systematically related to adult criminality. In order to contribute to a better understanding of the role of physiological factors operating in the longitudinal development of conduct, a series of cross-sectional analyses of the relations between various aspects of conduct and autonomic activity/reactivity were performed. As an important factor in the process, ability to concentrate was introduced in the analyses.

The Differential Relation of Aggressiveness and Restlessness to Autonomic Activity/Reactivity

The starting point for the cross-sectional analyses was the significant, negative relations that were obtained between aggressiveness and motor restlessness on the one hand, and adrenaline excretion on the other when the subjects were about 13 years of age. Boys rated high in aggressiveness and/or high in motor restlessness had a significantly lower excretion of adrenaline than boys rated low in these respects. The relation was particularly strong in the stressful situation. This was shown by the observation that those who were rated low in both aspects of conduct increased their adrenaline excretion from the nonstressful to the stressful situation more than boys who were rated high. This is of interest for further analyses.

In order to better understand the role of autonomic activity/reactivity in the process underlying various aspects of conduct, an analysis of the differential relation of aggressiveness independent of restlessness or restlessness independent of aggressiveness to low autonomic activity/reactivity is needed. This question was investigated by a linear partial regression analysis in which factors related to restlessness were extracted from aggressiveness and vice versa in order to investigate the relationship to adrenaline level. In the first column of Table 7.2, the partial correlation coefficients are presented for the relationship between aggressiveness in the absence of restlessness and adrenaline level, in the nonstressful and the stressful situation. In the second column the coefficients for the correlation between motor restlessness without aggressiveness and adrenaline level in the same situations are given.

The tendency is clear. When motor restlessness is removed from the ratings of

TABLE 7.2.
Partial Coefficients for the Correlation Between Aggressiveness in the Absence of Motor Restlessness and Adrenaline Excretion in a Nonstressful and a Stressful Situation, Respectively, and Between Motor Restlessness in the Absence of Aggressiveness and Adrenaline Excretion in the Same Situations.

	Aggressiveness without Motor Restlessness— Adrenaline Excretion	*Motor Restlessness without Aggressiveness— Adrenaline Excretion*	
Passive situation	.01	−.19	$p < .05$
Active situation	.00	−.22	$p < .05$

aggressiveness, the correlation between that which remains and the adrenaline level is zero. When aggressiveness is removed from the ratings of restlessness, the coefficients remain significant.

The results presented in Table 7.2 are illustrated in Figures 7.3a and 7.3b. The adrenaline level for those who were rated as highly restless without being highly aggressive is almost the same as for those who were rated as both highly aggressive and highly restless. This is consistent with the finding that those who were rated as highly aggressive without being highly restless, have about the same

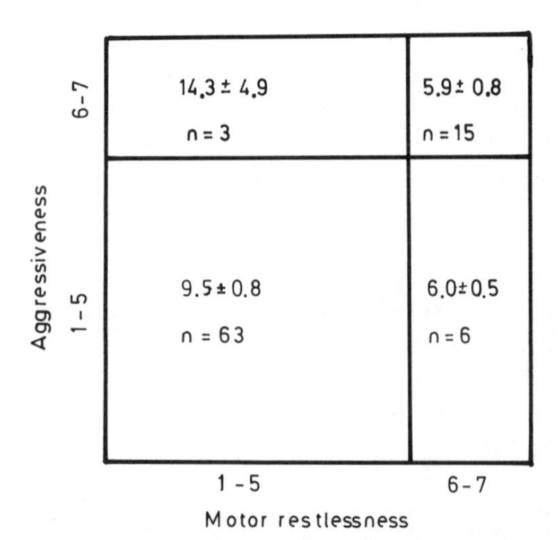

FIGURE 7.3a. Adrenaline excretion (ng/min) in a non-stressful situation at the age of 13 for individuals with different combinations of ratings of aggressiveness and motor restlessness at the same age.

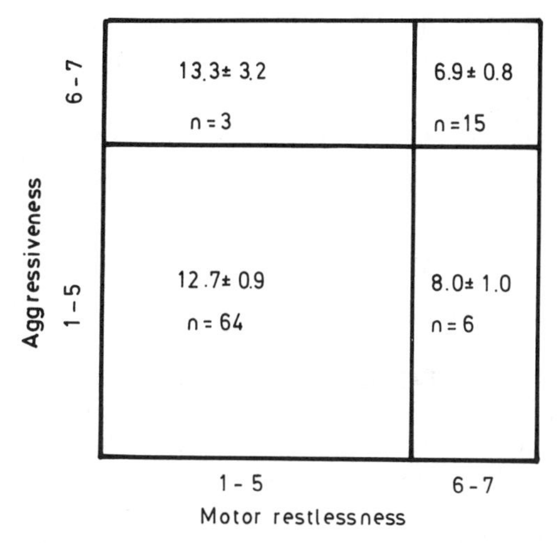

FIGURE 7.3b. Adrenaline excretion (ng/min) in a stressful situation at the age of 13 for individuals with different combinations of ratings of aggressiveness and motor restlessness at the same age.

adrenaline level as those who were neither restless nor aggressive. The illustrations in Figures 7.3a and 7.3b are based on a small number of individuals in some cells. However, it should be observed that the tendency is exactly the same under the two conditions. The pattern of relations in the total matrix of results for the relationships between adrenaline level at the age of 13, ratings of aggressiveness and of motor restlessness at age 13, and registered criminal offenses at adult age is very consistent.

Hyperactivity and Physiological Arousal in Terms of Adrenaline Excretion

As referred to above, the hyperactive syndrome is characterized by three main components: restlessness, lack of concentration, and impulsivity. The analyses presented in the preceding section were based on data for motor restlessness. Thirteen-year-old students' lack of concentration was also rated by the teachers. The importance of this aspect of hyperactivity is reflected in the definition of Attention Deficit Disorder with Hyperactivity (ADD-H), according to DSM III. Lack of ability to concentrate may be concerned with cognitive deficits. Douglas (1983, in press) argues that three aspects of lack of self-regulation appear to be the important elements in this deficit: organization and planning, mobilization and maintenance of effortful attention, and inhibition of inappropriate responding.

In one analysis of the differential relation of hyperactivity and aggressiveness to physiological arousal, in terms of adrenaline excretion, the pooled ratings of motor restlessness and lack of concentration were used as an indicator of hyperactivity. The findings confirmed the earlier results. Ratings of lack of concentration only also confirmed previous results, but the relationship was not as strong (Magnusson & af Klinteberg, 1986). These results led to a further analysis of the relation between motor restlessness and lack of concentration, as main components of hyperactivity, on the one hand, and adrenaline excretion on the other.

The Characteristics of Behavioral Hyperactives

When motor restlessness and lack of concentration are discussed as components of hyperactivity, it is necessary to observe that they cannot be used in isolation from each other as indicators of hyperactivity in the clinical sense. It is possible to be highly restless without corresponding to the clinical concept of motor restlessness. All teachers have seen boys or girls who appear restless, until they are given a task that really interests them and then concentrate on their subjects. The same is true for lack of concentration. Among those boys and girls who do not concentrate on their tasks at school are also those who lack motivation; that is, those whom teachers usually refer to as lazy. Thus, in order to be behaviorally hyperactive, an individual has to be characterized by both restlessness and lack of concentration.

The hypothesis that those who are characterized by both restlessness and lack of concentration can be distinguished as a group with respect to what Douglas (in press) discussed as "mobilization and maintenance of effortful attention," can be tested by using data from the mental test that was administered during the stressful situation. If the hypothesis is valid, the hyperactives in this sense should perform more poorly than those who are neither restless nor have concentration difficulties, and more poorly than those who are restless or have concentration difficulties (but not both).

The analysis was performed for (a) the number of items finished, (b) number of correctly solved items, and (c) the percentage of correct answers. The figures for the percentage of correctly solved items are presented in Table 7.3.

TABLE 7.3.
Percentage of Correctly Solved Items for Various Combinations of Motor Restlessness and Lack of Concentration

Motor Restlessness	*Lack of Concentration*	*Mean*	*Sd*	*N*
Low	Low	87.3	13.1	49
Low	High	82.9	11.7	14
High	Low	88.1	11.3	17
High	High	68.3	7.4	25

For all three measures boys who were high in both motor restlessness and lack of concentration had the lowest scores, as postulated by the hypothesis. An analysis of variance showed that the difference among the four subgroups were significant at the 5% level for the number of finished items, at the 1% level for the number of correctly solved items, and at the .01% level for the percentage of correctly solved items.

Because all three parameters tested might be influenced by individual differences in intelligence, a covariance analysis was performed in order to study the significance of the results. With intelligence controlled, the overall difference among the four groups was not significant for the number of finished tasks, but was significant at the 5% level for the number of correct answers and at the 1% level for the percentage of correctly solved items.

These results support the hypothesis that those who are characterized by both high restlessness *and* lack of concentration can be distinguished as a group. This distinction has not always been made in the published theoretical and empirical analyses. For the planning of effective empirical research and for the interpretation of results from such studies, it seems important to maintain this distinction.

Physiological Reactions to External Stimulation Among Behavioral Hyperactives

In order to understand the relationship of hyperactivity in terms of motor restlessness and lack of concentration to physiological arousal, a further analysis was performed. The difference between the "true" hyperactives and the others was of particular interest. In light of the difference in performance in the mental task, one aspect of particular interest is the differential reaction in physiological response to the external stimulation presented by the stressful situation.

In Figures 7.4a–b adrenaline excretion in the two situations is presented for four groups of boys: (a) those who were neither particularly restless nor particularly unconcentrated, (b) those who were restless without being unconcentrated, (c) those who were very unconcentrated without being particularly restless, and (d) those who were both very restless and very unconcentrated.

Figure 7.4a shows that in the non-stressful situation, the group of hyperactives (high-high), had the lowest excretion of adrenaline, as expected. The difference among the four subgroups was not significant. In the stressful situation, the hyperactives still showed the lowest excretion of adrenaline, and this was significant, at the 5% level. In a covariance correction for base line excretion, the adjusted difference in adrenaline output in the stressful situation between hyperactives and other subjects was significant at the 5% level ($F=3.93$). This implies that the other groups increased their adrenaline levels to a larger extent than did those who were characterized by both motor restlessness and lack of concentration. This can be seen in Figure 7.4a and 7.4b.

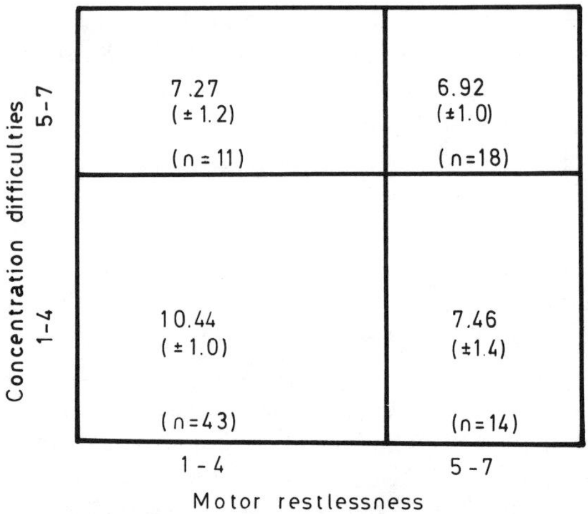

Concentration difficulties	Motor restlessness 1 - 4	Motor restlessness 5 - 7
5 - 7	7.27 (± 1.2) (n = 11)	6.92 (±1.0) (n=18)
1 - 4	10.44 (± 1.0) (n=43)	7.46 (±1.4) (n=14)

FIGURE 7.4a. Adrenaline excretion (ng/min) in a non-stressful situation at the age of 13 for boys with different combinations of ratings of concentration difficulties and motor restlessness.

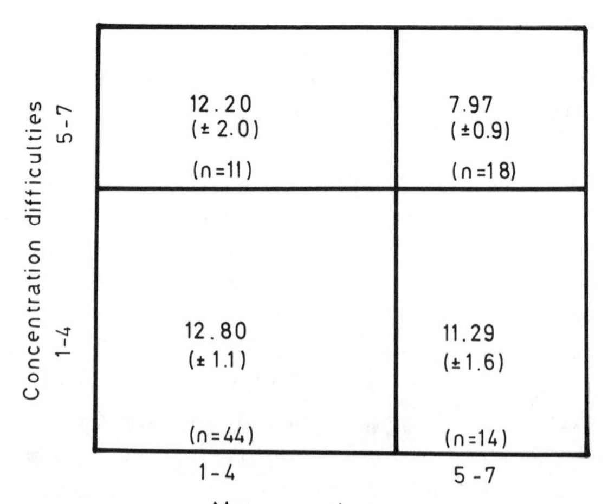

FIGURE 7.4b. Adrenaline excretion (ng/min) in a stressful situation at the age of 13 for boys with different combinations of ratings of concentration difficulties and motor restlessness.

In all three groups of boys, except those characterized by both high motor restlessness and high lack of concentration, there is a significant increase in adrenaline excretion from the nonstressful situation to the stressful. This is not the case for the "true" hyperactives. The mean increase in adrenaline excretion for them was 1.05 nanogram/min or 15% of the excretion under non-stressful conditions, while it was 3.1 nanogram, as an average, or 33% of the excretion under nonstressful conditions for the rest of the boys. The difference in increase of adrenaline excretion was significant at the 5% level.

The character of the two situations in which adrenaline excretion was studied should be kept in mind when interpreting the results. The nonstressful situation could not reasonably be perceived as threatening or demanding by anyone. However, the normal reaction to the stressful situation should be one of excitement and exertion, since it presented a challenge. Thus an interesting characteristic of the "true" hyperactives as defined here and one that may contribute to the understanding of their behavior is that physiologically they react significantly less to external stimulation than do others.

SUMMARY AND COMMENTS

The results presented in this chapter show a consistent picture. (It should be emphasized that the results presented here are in no case selected from a number of calculations on various issues. They were all the results of planned analyses.) Both aggressiveness and motor restlessness showed a systematic positive correlation with adult criminal activity as reflected in official registers. Particularly strong was the relation between the combined ratings of aggressiveness and motor restlessness at an early age and registered criminality at an adult age.

The hypothesized relation between high criminality at adult age and autonomic activity and reactivity at the age of 13 was also substantial; those who had been registered for criminal offences as adults had a mean level of adrenaline excretion at the age of 13 that was not only significantly lower but also considerably lower than the level for those who had not been registered. Thus the best predictor of criminal behavior is the combined aspects of conduct and physiological reactions at an early age. However, high predictions at that level of complexity are of little help in understanding the lawful principles of the developmental processes antecedent to adult maladjustment; just as little as the high predictions that can be made about differences in mean summer and winter temperature help us to understand the processes underlying change in the weather.

In order to contribute to the understanding and explanation of the processes in which early conduct is involved, a series of analyses were performed at a lower level of complexity, using data at a level that was consistent with the level of the processes, including situation specific data. At an early age both motor restlessness and aggressiveness showed a systematic, negative relationship to autonomic activity/reactivity in terms of adrenaline excretion. The partial regression analysis then demonstrated, in contradiction to earlier findings of research, that high aggressiveness per se was not systematically related to autonomic activity/reactivity in terms of adrenaline excretion. The analyses instead revealed a systematic relationship between hyperactivity per se, (in the absence of aggressiveness) in terms of motor restlessness and lack of concentration, and adrenaline excretion. Further analyses showed that those who were characterized by both high motor restlessness and high lack of concentration differed significantly from the others in their achievement on a mental task, that involved demand of maintained attention. They also deviated from "normal" in autonomic activity/reactivity in relation to external stimulation. These results lead to the hypothesis that the group of behaviorally hyperactives is responsible for the systematic negative correlations between conduct disorders and autonomic activity/reactivity in terms of adrenaline excretion that have been obtained. This hypothesis and the relationships to adult criminality and other types of maladjustment of mental conduct and the physiological aspects will be the object of further analyses in the project.

The series of studies discussed above have been reported here for illustrative purposes. Some of them have been published and others are under publication (Magnusson, 1987; Stattin & Magnusson, 1984). For details concerning data and the procedures and for extended discussions of the results, the reader is referred to these sources. Here five issues of particular interest for the purpose of this book are discussed.

Interpretation of the Results

In a traditional mechanistic model, the biological factor represented here by adrenaline excretion would be regarded as the cause of both conduct in a current perspective and adult delinquency in a longitudinal perspective. However, individual subsystems of cognitions, emotions, physiological factors, and conduct, are in a constant, reciprocal interaction and influence each other in current processes with consequences for all systems involved in a developmental perspective. During the developmental process both the psychological and the biological subsystems change, not only as a result of maturation, but also as a result of the interaction among the subsystems, that is, as a result of experiences in a wide sense. The character of this interaction among subsystems within the individual depends on the character of the environment with which the individual is interacting. Given this process, the interpretation of results in terms of cause and effect is not self-evident. It should also be observed that the simultaneous appearance of low adrenaline level and high aggressiveness and high motor restlessness at an early age among those later registered for crime may not necessarily imply that one of these factors causes the other(s) or even that they have a common cause (cf. Hofer, 1982).

Of critical importance for the understanding of the lawful processes underlying extrinsic maladjustment, were the results showing that those characterized by both high restlessness and high lack of concentration, react significantly less strongly physiologically to situational stress. Various interpretations have been offered. One interpretation, suggested by Satterfield and Dawson (1971), is that the restlessness exhibited by these children represents one way for them to achieve an increased and more satisfactory physiological level through self-induced increases in sensory stimulation. Another interpretation of interest for a discussion of the role of biological factors in the social development process can be made in terms of the concept of "fearlessness" (Lykken, 1982).

In this view the low adrenaline increase in some boys in the stressful situation may be interpreted as a low general reactivity to external stimulation in the autonomic physiological system. Then the feedback system, in which the individual's appraisal of information from the social environment about its positive vs. negative reactions to various behaviors, and the individual's interpretation of this information in terms of threats and demands do not function effectively in these children. This inefficiency may be either biologically determined from the beginning or the result of a lack of consistency in the environment's reactions to the child's early behavior. Therefore, it cannot play the important role as it does for other children in the interaction process through which the individual learns to deal with the external world. This reasoning leads to the question about the role of the perceptual-cognitive processes by which we interpret and evaluate the information from the external world, and if and to what extent deficiencies in that system contribute to disturbances in the socialization process. The results from the study using data from the Bender and Benton tests, briefly mentioned in the Prologue of this volume, support a line of thinking in this direction.

The Operating Factors

One of the cornerstones in the planning of the longitudinal project was the assumption that it is not possible to understand and explain the developmental process without considering the biological side of individual functioning and incorporating biological factors in the design of the study. The results presented in this chapter confirm the validity of that assumption for the understanding and explanation both of conduct in a current perspective and of the social developmental process.

To understand the role of biological factors in the processes underlying criminality, the result of preliminary analyses are of interest. The total group of males registered for crime before the age of 18 was subdivided in two groups: those who were only registered before the age of 20 and those who had also been registered later. The correlation between the number of crimes for each individual and adrenaline excretion at the age of 13 was then higher for those who were still active at adult age than for those who had conformed by that age. This contradicts the common assumption that the longer the interval between measures on factors involved in the development process, the lower the correlations. This result will be followed up in further analyses. One interpretation could be that biological factors are involved in the processes that lead to prolonged criminal activity. Another complementary interpretation is that low autonomic activity/reactivity as reflected in low adrenaline excretion, is related to a certain type of crime—a type that is more frequent among older criminals than among younger ones.

Person-Situation Interaction

A basic element in an interactional view on individual functioning is the assumption about the differential role of situational factors in the process of interplay among person and environment factors. This assumption is supported by the investigations of problem solving and physiological activity/reactivity in the nonstressful and the stressful situations. It was further verified in the project in analyses on physiological reactions among under- and over-achievers in the same two situations as reported by Bergman & Magnusson (1979). In the nonstressful situation no difference was observed between the two groups of boys, but the difference was substantial and significant in the stressful situation showing a higher adrenaline excretion for overachievers. (This result, seen in relation to the difference between behaviorally hyperactives and others in adrenaline excretion in the two situations, naturally leads to the question about a possible systematic relation between hyperactivity and underachievement. This question will be elucidated in further analyses.)

Level of Data Analyses

The results reported in this chapter clearly illustrate one of the main points discussed in Chapter 3, namely the importance of using data congruent with the level of the processes under consideration.

A common feature of personality research is that data are aggregated to enable the researchers to work with more reliable measures. This aggregation is often done without consideration of the complexity of the processes studied, as if aggregated data always should be preferable. Aggregation is often based on factor analyses, in which items that "go together" are used to form "homogeneous" factors. A factor analysis of the teacher's ratings in the project gave one common maladjustment factor in which aggressiveness and restlessness formed the core variables. (The ratings of these two variables had a correlation coefficient of .66.) That result could have motivated the aggregation of data for the two aspects of conduct to a composite measure; no one would have criticized such a procedure.

However, the analyses that have been presented illustrate the danger of remaining at the factor level independent of the problems at hand. From a purely statistical point of view, it is possible to show that factors usually regarded as homogeneous, when obtained by factor analysis, actually may contain items with rather low intercorrelations. The analyses presented, demonstrate the importance of combining studies at the factor level, when appropriate, with studies of more specific and well defined aspects of individual functioning in more finely-grained analyses in order to detect the functional relationships between important variables (cf. James, 1982).

The Validity of the Data

Because we can only understand and explain relationships among phenomena by studying the relations among observations, the fundamental requirement for meaningful and effective research on developmental processes, as well as in psychological research in general, is that high quality data are collected and used. The results presented in this chapter have been based on data from teachers' ratings, official registers, laboratory measures, and mental tests. The very consistent and psychologically-meaningful total picture that emerges from the results is a strong indication that the data are of a quality that permits us to contribute with solid knowledge to the understanding of the developmental background of adult functioning. In general, the very coherent picture of relationships among data for psychological and biological factors obtained in other studies in the project supports the use of the traditional methods for data collection in psychological research, as long as they are used in research that is carefully planned and implemented.

Further Analyses

The results presented in this chapter lead to new questions and further hypotheses about cross-sectional and longitudinal relations among person and environment factors involved in the developmental process. Some of these can be empirically investigated using the broad spectrum of data that are available.

The analyses presented here were concerned with the developmental antecedents of adult criminality. Then the analyses of criminality were based on the simple straightforward distinction between those who had been registered and those who had *not* been registered, independent of the number of crimes or the type of crime. Further analyses that are under way will naturally be more specific and use the information that is stored in registrations about number and types of crime, as reported in Chapter 5. With reference to the general theoretical framework for the project, and to the empirical results presented in Chapter 5 about the interdependence of the three main aspects of adult maladjustment, the analyses will also include data for psychiatric care and alcohol problems. They will also include data for personality factors that are relevant for understanding the processes underlying adult maladjustment, such as impulsivity, socialization, and monotony avoidance. A broad perspective is also important for understanding and explaining the role of early functioning, reflected in the interplay of mental factors, biological factors, and conduct as antecedents of adult maladjustment.

For further analyses, one important task is the study of the developmental pattern of mental processes, conduct, and physiology behind adult maladjustment for females. Females will have developmental differences if basic theoretical assumptions about the embeddedness of hormonal effects in the developmental context are correct. These assumptions are supported by preliminary analyses in the project.

Chapter 8

THE DEVELOPMENT OF PATTERNS IN ADJUSTMENT PROBLEMS: EARLY AGE TO ADULTHOOD

INTRODUCTION

The analyses reported in this chapter are logical consequences of the formulations of the theoretical framework for IDA presented in Chapters 1 and 2 and the methodological implications of such a framework discussed in Chapter 3. A basic proposition of these formulations was that individuals function as total organisms and that individuality is best reflected in partly specific configurations or patterns of aspects of individual functioning. One important consequence of this view is that research on individual differences as a basis for conclusions about the processes underlying individual functioning cannot be restricted to analyses of single variables, but has to include, as an important element in the research process, analyses of individuals in terms of their partly specific patterns of relevant factors. This chapter is devoted to the presentation of an empirical study using this perspective.

This view demonstrates the need for methods that adequately reflect individual differences in patterns of factors under investigation. In Chapter 3 cluster analysis and configuration frequency analysis were presented as methods that directly meet this need. One aim of this chapter is to demonstrate how these methods can be applied and have been applied in IDA for the study of developmental issues in terms of configurations of relevant factors involved.

The empirical analyses presented here concern the development of patterns of adjustment problems from an early age to adulthood. The multivariate description of development of adjustment problems in terms of patterns is reflected (a) in data from school, consisting of teachers' ratings of relevant conduct factors, data for underachievement based on standardized tests, and sociometric data for poor peer relations, and (b) in data from official records. The last type of data covers three main areas of maladjustment; criminal offenses, psychiatric problems, and alcohol abuse.

The factors that are investigated were chosen for the analyses due to the large amount of research that has been directed to the study of the relationship between person factors on the one hand and criminality, mental diseases, and alcohol problems on the other, both in a current and in a developmental perspective.

When the relationships between factors in childhood and adult adjustment have been investigated, factors in the environment (such as parents' divorce, parents' education and income, and parental behavior) and person factors (such as intelligence, conduct, achievement, and peer relations) at an early age have been traditionally treated as predictors, and criminal behavior, mental diseases, and alcohol problems as criteria. An important feature of this research is that it most often has been concerned with the pairwise relation between single variables on both sides (cf. Farrington, 1985). Almost independent of the choice of predictors and criteria, systematic and sometimes strong relations have been found between negative factors in the early environment and low intelligence, antisocial conduct, poor social relations, and poor achievement at an early age on the one hand, and criminal activity, mental problems, and alcohol abuse in adulthood on the other.

One main question arises. To what extent does the total picture of pairwise correlations depend on a rather limited number of persons characterized by broad syndromes of adjustment problems, in that they may generate the relationships that appear in studies on each of the single aspects that together constitute the syndromes? A consecutive question follows naturally: To what extent is it meaningful to interpret correlations between single conduct problems at an early age and indicators of adult maladjustment as saying something about the significance of specific early behaviors in the development process if they usually occur and have long-term consequences only in syndromes of several adjustment problems (Morris, Escoll, & Wexter, 1956)? It was one of the aims of the analyses presented here to contribute to the elucidation of this issue, which has important implications for theory as well as for empirical research on the developmental background of adult functioning.

The empirical analyses in this chapter are presented for illustrative purposes. The finely graded analyses are under way and further analyses are being planned.

METHOD AND DATA

Samples

The study is based on data for males from the main group; the cohort that has been followed since the age of 10. This is the same group for which the longitu-

TABLE 8.1.
Group Size and Drop Out. (Main Group, Males).

Type of Data	*Group Investigated*	*Drop Out*	*Total Cohort*
a) School data at age 13	540	5	545
b) Data from official records	705	3	708
c) Data from a + b	538	7	545

dinal analysis of the relation between aggressiveness and motor restlessness at the age of 13 and adult criminality was presented in Chapter 7.

This particular study is based on data concerning conduct problems, poor social relations, and underachievement all collected at the age of 13. Official record data for criminal offenses, mental problems, and alcohol abuse cover the age span from birth through the age of 23, with a division of the data in 2 groups, for the age span 0 through 17, and the age span 18 through 23. Data from official records were collected for males from the total main group; all children entering the cohort before the end of grade 9 were included.

The group size and the drop out figures for the various types of data are reported in Table 8.1. The group with school data from age 13 consisted of all children in grade 6 for which the necessary information about adjustment problems was available. Data of this kind were missing for 5 boys, (about 1% of the complete grade 6 cohort.) As the collection of data from official records is complete due to the all-encompassing Swedish population register system, the only dropout that occurred was for those who died or emigrated. When studying adjustment problems, the representativeness of the sample is of decisive importance for the generalization of results, since persons with adjustment problems tend to be over-represented in the dropout (Bergman, Hanve, & Rapp, 1978; Bergman & Magnusson, 1987; Cox, Rutter, Yule, & Quinton, 1977).

Adjustment Data from the Age of 13

At the age of 13, the boys were grouped into clusters of similar patterns on the basis of the following 6 indicators of extrinsic adjustment problems:

Aggressiveness
Motor restlessness
Lack of concentration
Low school motivation
Underachievement
Poor peer relations

The first four aspects of adjustment were covered by teachers' ratings performed on 7-point scales for each variable (see pp. 154–155). Underachievement was measured for each individual as the difference between his obtained achievement score and the predicted score from his intelligence level. Poor peer relations were based on sociometric ratings and primarily measured unpopularity.

An optimal investigation of individual patterns would imply the use of some sort of absolute measures of the variables under investigation. This is a difficult prerequisite for many relevant variables in psychological research. In order to meet the requirement as far as possible, the analyses reported here were based on data for what Bergman and Magnusson (1983) designated as *semi-absolute* scales. Using that approach each of the indicators just mentioned was scored on a 4-point semi-absolute scale with the following tentative definitions of the adjustment scores.

Adjustment score	*Description*
0	*No adjustment problem*
1	*The adjustment is not good but no clear problem*
2	*An adjustment problem*
3	*A pronounced adjustment problem*

For instance, with regard to the teachers' ratings, which were made on 7-point scales, the values 1–4 were coded 0, 5 was coded 1, 6 was coded 2, and 7 was coded 3. The reasons for this scaling were discussed in Bergman and Magnusson (1983, 1984a, 1987). In the first of these references a more detailed presentation of the indicators and the underlying theoretical background and taxonomy were given.

Pattern Analysis on Data from the Age of 13

For the grouping of the children, a cluster analytic procedure—RESCLUS—developed in the project was applied (Bergman, 1985; Bergman & Magnusson, 1984a). The individuals were first grouped according to a hierarchical method. This classification was then taken as input for a relocation cluster analysis. One special feature of the relocation procedure employed is that a residue of unclassified individuals are formed. This is important for two interrelated reasons: First, it is important from a theoretical viewpoint to recognize the fact that not all individuals fit well into one of a small number of classes or types. Second, deficient reliability in data will yield profiles for some individuals that do not fit well into any of the categories. The final clustering of the boys based on data from the age of 13 is given in Table 8.2.

According to the solution presented in Table 8.2, most boys (296 out of 540) belong to a cluster characterized by no adjustment problems. For the presenta-

TABLE 8.2.
Final Clustering Solution for Extrinsic Adjustment Problems.
Boys at the Age of 13.

			Cluster Means					
Cluster No	*Size*	*Average Coefficient*	*Aggressiveness*	*Motor Restlessness*	*Lack of Concentration*	*Low School Motivation*	*Under-achievement*	*Poor Peer Relations*
1	296	.12	—	—	—	—	—	—
2	23	.30	—	—	—	—	—	2.4
3	40	.28	—	—	—	—	2.6	—
4	61	.39	1.3	1.4	—	—	—	—
5	41	.39	—	1.5	2.3	1.9	—	—
6	12	.56	1.7	1.8	2.3	1.9	2.6	—
7	37	.37	2.3	2.3	1.9	1.3	—	—
8	22	.48	2.2	2.7	2.6	2.4	—	1.9
Residue	8		1.5	1.4	2.3	1.3	1.3	2.3

Note. — means that the cluster mean of a variable is less than 1.0 in the 4-point scale coded 0, 1, 2, 3. Average coefficient means average error sum of squares within the cluster.

tion here, some characteristic maladjustment clusters are of particular interest. Three multiproblem clusters, indicating severe problems, are identified (6, 7, and 8). The most severe is cluster 8, which is characterized by strong maladjustment symptoms at the age of 13 for all the factors investigated except underachievement. Cluster 7 is characterized primarily by high aggressiveness, motor restlessness and, lack of concentration. The strongest indicators of maladjustment in cluster 6 are underachievement, lack of concentration, and low school motivation, which might be interpreted as indicating learning problems. It is interesting to note that when only aggressiveness and motor restlessness form a cluster together (cluster 4), the indications of maladjustment are only very remote. Of particular importance for the problems formulated in the introduction about the study of single variables versus patterns is the observation that both poor peer relations and underachievement, respectively, characterize single-problem clusters at the same time as each of them is one of several symptoms in multiproblem clusters. As shown by Bergman and Magnusson (1984b), these two single-problem clusters were identified also at the age of 10 and showed a significant stability from 10 to 13. It is also worth noting that none of the other single variables, for example aggressiveness, constitute distinctive clusters themselves. These findings have important implications and will be further commented upon.

Cross-validation of Patterns. A crucial question for the interpretation of results obtained from cluster analysis is the generalizability of the results, i.e., the extent to which the obtained patterns are dependent on sample characteristics. An estimation of the ability to generalize from the results obtained with the clustering method was made in an earlier study in which a pattern analysis was performed on data from the age of 10, for the same variables as presented here (Bergman & Magnusson, 1984b). The total sample was split into two random halves and the same procedure was used for clustering the individuals in the two subsamples. The solutions showed a satisfactory degree of similarity and there is no reason to assume that the same would not hold for the sample studied here.

Adjustment Data from Official Records

The presence of criminal offenses, psychiatric care, and alcohol abuse in official records were identified for the first age period (10–17 years of age) and the second age period (18–23 years of age).

For adults, criminal offenses were measured as those that resulted in court decisions. For children below the age of criminal responsibility (15 years), the number of confirmed crimes was counted if, after a police investigation to determine guilt, the child was turned over to the welfare authorities and investigated by them. In this analysis, alcohol offenses and minor traffic offenses were not counted in order to avoid contamination with records for alcohol abuse. (For further details, see Stattin, Magnusson, & Reichel, 1986).

Psychiatric care was defined as having had an interview with a psychiatrist that led to a DSM III diagnosis. A pure alcohol diagnosis was not counted as psychiatric care but instead as severe alcohol abuse. A pure drug diagnosis other than alcohol was not counted at all (only 2 persons were thus categorized). All psychiatric hospitals in the vicinity of the subjects' homes and all general hospitals with psychiatric clinics were searched for psychiatric records (see von Knorring, Andersson, & Magnusson, 1987). Alcohol abuse was measured by counting the number of registrations of alcohol offenses such as being taken in by the police for drunkenness, etc.

Pattern Analysis of Data from Official Records. The pattern analysis of data from official records was performed with configural frequency analysis (CFA). This type of analysis tests whether an observed value pattern occurs significantly more often (a *type*) or significantly less often (an *antitype*) than expected by an independence model. In the testing of the significance of the findings, the exact binomial probability model was used to avoid the problems with small expected values (Bergman, 1985). The significances were corrected for by the fact that several simultaneous and dependent tests were performed, according to the so-called Bonferroni method (see Krauth & Lienert, 1982). In Table 8.3 the results are given of configural frequency analyses (CFA) of official records data.

The results presented in Table 8.3 show three significant types at both age periods: (1) Not to be in the official records at all, (2) to be in the official records for both criminality and alcohol abuse, and (3) to be in all three kinds of records. Thus, the frequencies in these three categories are significantly and much higher than could be expected from a random model. The category characterized by having both a criminal and an alcoholic registration actually contains about

TABLE 8.3.
Cross-Sectional Configurations of Data from Records for Criminal Acts, Psychiatric Care and Alcohol Abuse.

			Age 0–17		*Age 18–23*	
Criminality	*Psychiatric Care*	*Alcohol Abuse*	*Obs Freq*	*Exp Freq*	*Obs Freq*	*Exp Freq*
No	No	No	500*t*	463.41	504*t*	443.10
No	No	Yes	12*at*	34.61	22*at*	66.50
No	Yes	No	29	39.99	25*at*	41.22
No	Yes	Yes	0	2.99	6	6.19
Yes	No	No	109*at*	140.48	71*at*	117.73
Yes	No	Yes	28*t*	10.49	48*t*	17.67
Yes	Yes	No	18	12.12	13	10.95
Yes	Yes	Yes	9*t*	0.91	16*t*	1.64

t = type, significant at the 5 per cent level
at = antitype, significant at the 5 per cent level

three times as many males as expected, and the category characterized by all three kinds of registers contains about ten times as many as expected. This is true for both age levels. On the other hand it is interesting to note that being in only one of the three types of official records is a significant *antitype* (except for being in psychiatric care at age 0–17), i.e., it occurs significantly less frequently than expected if the occurrences were randomly distributed in the sample. Being registered only for alcohol problems occurs only about one third as often as expected.

The results presented in Table 8.3 reflect the same picture as was discussed in Chapter 5 with reference to Table 5.17. The analyses presented there showed that more than half of those registered at adult age were found in more than one register. The result of these analyses has relevance for further discussions about variable- versus pattern-oriented research on maladjustment.

LONGITUDINAL ANALYSES

Adjustment Patterns at Age 13 as Related to Adult Maladjustment

In Table 8.4 the longitudinal relationship between school adjustment clusters at age 13 and adjustment problems at age 18–23 are reported as reflected in criminal, psychiatric, and alcohol records. The tabled information reflects the percentages that appear in records for criminal acts, psychiatric care, and alcohol abuse, respectively. The significance testing of types and antitypes in the relationships was performed using exact hypergeometric tests (see Bergman & El-Khouri, 1986).

Some interesting tendencies emerge in Table 8.4 with respect to the developmental background of adult criminality, mental illness, and alcohol abuse. Of particular interest for the elucidation of the developmental process underlying adult maladjustment are the significant types and antitypes that appear. Four significant types appear; those who were characterized by the patterns for the severe, multiproblem clusters 7 and 8 at the age of 13 (see Table 8.2) were significantly more often registered for criminality and/or alcohol abuse at adulthood than expected by chance. It is worth noting that neither the single-problem cluster characterized by underachievement or poor peer relations, respectively, or the multiproblem cluster characterized solely by weak aggressiveness and weak motor restlessness at age 13 were significantly related to being registered at adulthood. There is not even a conspicious tendency in that direction.

In concordance with the results summarized above, significant antitypes appeared for those who belonged to the group without problem indicators at age 13 and were being registered for criminality or alcohol abuse at adulthood. This

TABLE 8.4.
Relation Between Adjustment Clusters at Age 13 According to School Data, and Criminality, Psychiatric Care, and Alcohol Abuse at Age 18–23. (Males, *N* = 538).

Cluster at Age 13	*N*	*Percentages that Appear in Records for*		
		Criminality	*Psych Care*	*Alcohol Abuse*
1. No problems	295	13.6*at*	6.4	7.5*at*
2. Poor peer relations	23	13.0	13.0	4.3
3. Underachievement	40	10.0	5.0	12.5
4. Weak aggression, weak motor restlessness	61	18.0	0.0	13.1
5. Lack of concentration, low motivation	40	30.0	15.0	27.5
6. Multiproblem-severe underachievement and lack of concentration	12	33.3	0.0	8.3
7. Hyperactivity with aggressiveness	37	48.6*t*	16.2	37.8*t*
8. Severe multiproblems	22	50.0*t*	13.6	36.4*t*
Residue	8	25.0	0.0	12.5

t = problem *type,* significant at the 5 percent level
at = *antitype,* significant at the 5 percent level

implies that those who were categorized as having no conduct problems at the age of 13 appear significantly less often than expected from a random model in registers for criminal acts or in registers for alcohol problems.

A strong relation with adjustment clusters at age 13 comes up for *criminal records*. It is then interesting to note that for the clusters 2, 3, and 4, the percentage of those who have criminal records does not deviate to any marked extent from what is the case for males from the "no problem" cluster. It is when we come to the real multiproblem clusters 5, 6, 7, and 8, that the strong relationships appear, with the strongest one for clusters 7 and 8, in which about half of those who belong to these clusters appear in criminal records at age 18–23.

The correlation with adjustment clusters at school age is much weaker for *psychiatric care* than for criminality at age 18–23. For the clusters 3, 4, and 6, the percentage of males having a psychiatric record is actually smaller than the percentage for those who belonged to the "no problem" cluster. In two of these three rather large clusters (clusters 4 and 6), no male has a psychiatric record. The strongest relation appears for clusters 2, 5, 7, and 8. However, the percentages for those having a psychiatric record in these four clusters are not very high.

For *alcohol abuse* at age 18–23 there is a strong relation to three clusters at school age; clusters 5, 7, and 8. As emphasized earlier, these clusters at 13, also criminal records at age 18–23 show a strong relation. For both criminal and alcohol records the strongest correlation appears for cluster 7, in which the boys were characterized primarily by high aggressiveness, high motor restlessness, and strong lack of concentration, and for cluster 8, in which the boys were characterized by all problems included in the analyses except underachievement.

Thus, the total picture is that there is a strong correlation between syndromes of maladjustment in terms of behavior ratings at age 13 and adult criminal offenses and alcohol abuse as reflected in official records. Chi-square tests for each type of official records separately showed that the overall relation between categorization in terms of ratings at age 13 and categorization on the basis of official records at age 18–23 was significant at the 0.1% level for criminal and alcohol records, and at the 5% level for records for psychiatric care. There is a clear tendency that the relation is weaker between a maladjustment syndrome at the age of 13 and adult psychiatric care than between maladjustment syndromes and registration for criminal acts or alcohol problems.

It is of interest to note that there is no significant association with adult maladjustment, as reflected in official records, and the single-problem clusters of poor peer relations or underachievement at age 13. At first glance, this result may appear contradictory to the sizable relationships often reported between these variables and adult maladjustment of various kinds. However, viewed from the pattern approach perspective, there is no contradiction. Subjects with several conduct problems as the core symptoms appear to be responsible for the often observed relationships. For other subjects not characterized by this syndrome but who are characterized by only poor peer relations or underachievement, no clear relationship to adult maladjustment appears. As noted in an earlier section of this chapter, both poor peer relations and underachievement have been shown to constitute significant problem types, both cross-sectionally and longitudinally (from 10 to 13 years of age).

Do Early Official Records and Ratings of Conduct Problems Reflect the Same Antecedents of Adult Problems?

A question of particular interest for the discussion of implications of early indicators of maladjustment in adults is the following: Are early adjustment problems, as reflected in behavior at school and in official records for early maladjustment only different facets of the same generalized maladjustment syndrome or should they be viewed as expressions of connected but separate adjustment syndromes? This question was elucidated by calculations, presented in Table 8.5.

In Table 8.5 all subjects are grouped on the basis of two categorizations: if they had a criminal record or not during the age of 0–17 and if they belonged to

TABLE 8.5.
Criminal and Alcohol Records at Age 18–23 for Various Combinations of Belongingness to a Severe Adjustment Cluster (7 or 8) at Age 13 and Having Criminal Records at the Age 0–17

		Criminal Records at Age 0–17	
		Yes	*No*
Belongs to severe adjustment clusters at age 13	Yes	16/30	2/29
	No	19/88	12/391

one of the two most severe behavior clusters (clusters 7 and 8) at age 13. Each cell shows how many out of the total number of males in the cell have both criminal and alcohol records at age 18–23.

The main tendencies in Table 8.5 are obvious. The first and primary one is that it is the combination of belonging to a severe adjustment cluster at the age of 13 *and* having criminal records during the period 0–17 that is associated with a high risk for having a criminal and an alcohol record during the period 18–23 (53.3%). The risk is substantial, but much lower for those males who had only early criminal records but did not belong to a severe adjustment cluster at the age of 13 (21.6%). An important result with far-reaching theoretical and practical consequences is that those who belong to the severe maladjustment cluster at the age of 13 but do not have an early criminal record, do not show any sizeable increase in the combination of adult criminal and alcohol records compared to those who did not appear in severe adjustment clusters.

A configural frequency analysis with the table design corresponding to three dichotomous variables shows that the two cells in which persons appear who had a criminal record at the age of 0–17, are significantly overrepresented at the 5% level with respect to criminal and alcohol registrations at the age of 18–23. Thus, the results indicate that the relation between early criminal records and adult criminal and alcohol records is stronger than the relationship between early antisocial syndromes alone and adult records. Two methodological circumstances that might influence the size of the obtained relations should be considered. First, the relationship between early and adult records was studied in terms of the same type of data, while the relationship between early antisocial behavior syndromes and adult records was studied in terms of different types of data. Second, the ratings underlying the antisocial behavior syndromes were based on observations during the age period of 10 to 13 and performed at the age of 13, while the early criminal records covered a period that included a later age level, closer to the age period to which the adult records for criminality and alcohol problems generally refer.

COMMENTS AND CONCLUSIONS

In this chapter, longitudinal analyses of maladjustment in terms of behavior patterns at the age of 13 and registered criminal offenses, psychiatric care, and alcohol abuse have been reported. The study originated from the general framework of the longitudinal project. The results were presented as an illustration of the importance and fruitfulness of supplementing, in final analyses, the study of variable-directed, pairwise relationships with person-directed analyses in terms of patterns. They lead to a series of questions which will be elucidated in further theoretical and empirical analyses within the project.

The careful reader has observed that the main longitudinal analyses presented in this chapter were concerned with the correlation between behavior patterns at the age of 13 and each of the aspects of adult maladjustment covered by official records for criminality, psychiatric problems, and alcohol abuse. This analysis implies, that we have only marginally investigated the relationship between belonging to an adjustment pattern at the age of 13 and various configurations of types of maladjustment at adult age. However, the results of the pattern analysis done separately for criminal, psychiatric, and alcohol records have been presented here, since they contain information that is of interest per se, and since they illustrate the basis for further, more finely-grained analyses including analyses of patterns of adult maladjustment. Such analyses will be extended to include data also for females.

The illustrative character of the study implies that many theoretical, conceptual, and methodological problems connected with the general approach have not been discussed. Only a few comments will be made on the results and possible directions for further research will be suggested.

The Importance of Description

In Chapter 2 it was argued that careful observation and description of the phenomena under study guided by theoretical considerations about the relevance of specific data and the appropriate methods for observation and data treatment constitute the important basis for sound and effective theorizing. The study presented in this chapter is a good illustration of this proposition. It is basically a description of development in terms of patterns of adjustment from the early teenage period to adulthood. The description in such terms can offer the basis for further theorizing as well as for further empirical research on the developmental background of adult functioning.

The Significance of a Pattern Approach

The findings presented and discussed in this chapter bring into focus the possible problem associated with the study of pairwise relationships and the potential offered by a pattern approach, a discussion that was raised in the introduction to the chapter. The study of poor peer relations and underachievement as antecedents of adult maladjustment can serve as an illustration.

The two groupings based on poor peer relations and underachievement, respectively, appear as significant types at an early age. It is thus worth noting that membership in one of these two groups in itself does not entail increased risk of later maladjustment, a finding that has important implications for the interpretation of these constructs in relation to adult maladjustment. With regard to poor peer relations, for example, the results indicate the existence of at least two distinct groups; one group with only this problem and the other group with a general syndrome of adjustment problems, including severe conduct problems. This result gives quite another interpretation to the relationship between poor peer relations and later maladjustment than that given by, for instance, a correlation coefficient between an indicator of poor peer relations and an indicator of later maladjustment. It also appears very doubtful that even a more sophisticated variable analysis where, for instance, some indicators of conduct problems were controlled for before studying the correlation, would reveal the structure found in the pattern analysis.

The above illustrates that the higher-order interaction structure found in pattern analysis cannot always be reduced to pairwise relationships as is often done in variable-oriented analyses. It also indicates that observed relationships in a set of adjustment variables (in a cross-sectional as well as in a longitudinal perspective) may only be an expression of the fact that a small group of subjects are characterized by a general syndrome of problems. If the existence of such a group was controlled for, then significant relationships among the variables may disappear.

Preliminary analyses in the project support this assumption. When the subjects that were characterized by several severe adjustment problems (clusters 6, 7, and 8) at age 13, were controlled for in regression analyses of the relationship between single early behaviors and adult adjustment, the systematic relation of aggressiveness to adult criminality and to adult alcohol abuse, disappeared completely. Thus, aggressiveness, when not combined with other adjustment problems, does not seem to be the important antecedent as is assumed by theorists and supported by empirical research in the variable directed tradition. This, however, did not happen for some other conduct variables that were analyzed in the same way. In another preliminary analysis, the longitudinal stability of criminal activity and alcohol abuse was studied after the subjects with significant multiproblem configurations were removed. Criminal activity maintained a significant stability but alcohol abuse at age 0–17 was not related to alcohol abuse at age 18–23. An essential task for further research is to investigate if and to what extent stability and prediction coefficients can be explained by the occurrence of a small group of multiproblem boys who are stable in their behavior across time and to what extent single factors have an impact per se, independent of the fact, whether they occur together with other factors or not. The results of such studies will contain important information and contribute to a better understanding of the developmental process underlying adult functioning.

The variable orientation of traditional development research is reflected in research on two main issues. The first is concerned with the stability of single aspects of individual functioning, as discussed in Chapter 3, either because it is of interest per se or in order to elucidate the issue of personality consistency. Results from such studies have been used as arguments by both sides (cf. Mischel, 1968; Olweus, 1979). The second area has been concerned with the prediction of adult criminality, mental illness, and/or alcohol problems from single symptoms of maladjustment at an early age, for example, aggressiveness and hyperactivity. The empirical studies presented in this chapter have relevance for the interpretation of the empirical results reported on these two issues in the variable-directed tradition. One implication is that much of the magnitude of the stability coefficients and the coefficients for prediction of adult functioning from single maladjustment symptoms at an early age can be accounted for by the fact that the males who are characterized by several problem indicators at an early age are stable in their patterns of maladjustment through the life course and up until adulthood. There is continuity and stability in the patterns of individual functioning: It is individuals who are stable across time, not variables.

Antisocial Behavior Syndromes as Antecedents of Adult Maladjustment

An interesting result of our research is that even belonging to a severe adjustment cluster at the age of 13 is not an indicator in itself of adult criminality and alcohol abuse as they are reflected in official records. It is only when early misconduct appears together with a criminal record before 18 that there is a high risk for being registered for criminal offenses and alcohol abuse. This result again underlines the importance of one of the main propositions of the theoretical framework for the project presented and discussed in Chapters 1 and 2, that the individual is functioning as a totality and thus, that we have to study a broad perspective of aspects of individual functioning in order to truly understand the individual developmental process.

The finding that it is only in combination with having been registered before the age of 18 that even a severe syndrome of maladjustment at age 13 indicates a strong antecedence of adult maladjustment raises interesting questions: Can those who belong to a severe maladjustment cluster at the age of 13 be separated into two groups, one conforming early and not proceeding to further delinquency and alcohol abuse, and one group that continues into maladaptive course and appears in the records before 18? Or is it that the main difference between those who have been registered and those who have not been registered during the period of 18–23, among those who belonged to a severe adjustment cluster at age

13, is that some of them for various reasons have been registered early, with a lasting effect on their future behavior and on their risk of being registered while others have not?

Methods for Treatment of Data

The methodology applied in the analyses, the results of which have been reported in this chapter, has an interesting relationship to what Cattell (1965) designated Q- and P-techniques. In Q-technique, persons instead of variables are grouped as in some methods suggested here, although factors analysis is used as the technical tool. In P-technique the focus is on the individual's whole pattern of responses as in the person approach. However, the P-technique design is built basically for analyses of the single individual using data for many variables across many occasions. It is then replicated for a small sample of individuals (Nesselroade & Ford, 1985).

It is fully possible to use P-technique for carrying through a person approach. Whether this technique or the techniques suggested here are to be used naturally depends on the problem under study and available data. In the longitudinal project presented here, data are not available for many occasions and at least partly different variables are studied at the different occasions rendering applications of P-technique difficult.

In the preceding discussion the results have been interpreted so as to support the importance of supplementing the study of single aspects of individual functioning with studies of individuals in terms of their characteristic patterns of relevant aspects, and supplementing the regression models and methods for treatment of data with the types of methods used in the analyses presented here. As emphasized in the earlier discussion, configural models cannot be a substitute for regression models for all purposes in research on individual differences. The two approaches complement each other as demonstrated in the empirical studies presented in the foregoing and this chapter. One basic problem in regression analyses is the appropriate variables to consider in the equations. Hence, an important contribution of a pattern approach may be to provide information about the relevant variables to be introduced in such analyses and how they should be introduced. One central issue in the further analyses that are now being planned will be to investigate more carefully to what extent the two approaches can provide the same information and basis for possible explanations of the structures and processes under investigation and under what conditions their special merits have to be considered and used.

EPILOGUE

In the prologue to this book I described the personal background to the planning and implementation of the longitudinal project Individual Development and Adjustment (IDA) from which a series of volumes will be published under the title "Paths Through Life."

The empirical studies presented in Chapters 5 through 8 demonstrate from various perspectives the necessity of conducting longitudinal research in order to understand and explain the processes underlying individual development, as well as the role of individual and environmental factors in those processes. The importance of researching a broad set of aspects of individual functioning (including cognitive-affective factors, biological factors, and conduct), and illuminating the problems involved in micro- and macro-level developmental research was also shown in the studies reported. The empirical characteristics of the main cohort given in Chapter 5 demonstrated the necessity of studying development (both in children and adults) in terms of individual patterns of relevant factors. The usefulness of such an approach was shown in Chapter 8, in which appropriate methodologies for the study of patterns of development were applied and discussed. The essential role of biological factors in the development process was particularly illustrated in Chapters 6 and 7. The studies reported in Part II demonstrate the importance of a broad research strategy and methodological approach to developmental research in which cross-sectional and longitudinal studies, situation- and nonsituation-specific data (from extensive and intensive studies), and finely grained analyses at the micro- and macro-levels of the processes complement and support each other.

The methodological problems connected with the person approach that forms the theoretical frame for IDA are multifaceted. There is no single solution to these problems; the methodological solutions must also be multifaceted. Thus, the methodology that was presented in Chapter 8 for pattern analysis is only one possible way of handling some of the problems. For example, the problem of the existence of interindividual differences in growth rate, which affects cross-sectional data for many central factors involved in the developmental process (see pp. 68–74), still remains. The handling of this problem requires other solutions than those presented here, and the development of appropriate methods is a task for further research.

The empirical studies represent the first step in a more general plan for the use of the data from the longitudinal project in future analyses. Each presentation raises a series of new questions that can be addressed using data from the project. The study of biological maturation and its possible long-term effects will be extended to cover aspects of adult life not reported in Chapter 6 and analyses of the male cohort. The interplay of cognitive-affective factors, biological factors, and conduct as antecedents of adult functioning will be investigated by the inclusion of various aspects of individual functioning not reported in Chapter 7 and by more detailed analyses of types of adult criminal offences, mental problems, and alcohol abuse.

Studying developmental issues in terms of patterns of relevant variables has two distinct advantages. First, it allows for the important, sometimes very complex interactions that exist among factors operating in the developmental process. Second, the results and the generalizations based on them refer directly to individuals. Generalizations about individuals are the ultimate goal of our research endeavours. Chapter 8 demonstrates one possible approach to pattern analysis. The data used for the studies presented therein were restricted to data for various aspects of extrinsic adjustment. Further pattern analyses will naturally include aspects of intrinsic adjustment (such as satisfaction, sense of control, self-perceptions, and evaluations), biological factors and environmental variables. An important type of pattern analysis that is under way in the project uses individual developmental patterns for the variables under consideration as the basis for grouping the individuals into clusters that are homogeneous with respect to their general course of development. Further theoretical, conceptual, and methodological work is needed to refine the methodology so as to make it an effective tool. One central problem is to decide which are the "relevant" variables for analysis. To some extent they are chosen as a result of theoretical considerations. In other cases, the choice can be made on the basis of results from empirical analyses of single variables such as those presented in Chapter 7.

Some other essential areas, which are interesting in a developmental perspective where longitudinal research is urgently needed and for which IDA was partly planned, should be briefly mentioned. For example, an important dimension for understanding an individual's course of development from childhood to adulthood concerns choice, preparation, and involvement in a vocational career. The study of the factors operating in this process (e.g. structure and quality of family life during upbringing, individual potentialities and limits, and chance events) and the investigation of the role of a vocational career in the individual's intrinsic and extrinsic adjustment are necessary to understand the process underlying adult adjustment. This topic has been given a central role in the planning of IDA and has formed an important basis for the choice of information about the participants and characteristics of the environment covered by the data collected. Some empirical studies have already been presented using data from the initial and main cohorts (Dunér, 1978a,b; Ekehammar, 1977a,b, 1978a,b). Plans for further research in IDA naturally include studies on the issue of vocational career.

The investigation of some aspects of the participants' educational and vocational careers are, of course, affected by cohort effects; therefore, such effects should be studied. The general design of IDA enables the study of cohort effects in further research on the educational-vocational career process.

A connected area of research is that of the work environment as a determinant of physical and mental health. This issue has raised considerable interest during the last decade; much empirical research has been devoted to the topic. The overwhelming majority of the studies have been performed cross-sectionally. However, the limitations of the cross-sectional approach have become more and more obvious. There is, for example, the need to know the characteristics of individuals *before* they enter a given work environment in order to enable the separation of environmental effects from prior personal dispositions. IDA has resolved this problem by following personal and environmental factors in a large representative sample from childhood to adulthood. Thus IDA offers good opportunities for further research in this field.

Research on work environments is one element within a broader perspective that is concerned with the role of general upbringing and living conditions and general life styles for physical and mental health (Hamburg, Elliott, & Parron, 1982). The requirements for research on work environments are also valid here. Thus, IDA offers good opportunities for research on the developmental background of adult health. One example of such a study using data from IDA on Type A-behavior in a longitudinal perspective, from the age of 13 to 26, was presented by Bergman and Magnusson (1986). The results of such studies are of interest for understanding the developmental process and provide a basis for measures to contribute to individual health in a broader sense. As far as possible this line of research will be followed in the planning of future studies using IDA.

Finally, in the perspective of life span development, the type of strategy applied in IDA, in which a representative group of individuals is followed from an early age with respect to a broad set of personal and environmental variables, forms the most effective and sometimes necessary basis for an effective study of the personal and environment factors involved in the process of aging. (The oldest cohort, about 15 years of age at the time of the first data collection, is now 36.)

The effective analysis of these issues as an object of research in IDA presupposes much theoretical and methodological work. The substantive results of analyses of the project data and the theoretical and methodological considerations that will emanate from the project will be presented in forthcoming books in the series.

REFERENCES

Abramson, L. Y., Seligman, M. E. P., & Teasdale, J. O. (1978). Learned helplessness in humans: Critique and reformulation. *Journal of Abnormal Psychology, 87*, 49–74.

Achenbach, T. M. (1985). *Assessment and taxonomy of child and adolescent psychopathology.* Beverly Hills, CA.: Sage Publications.

Ainsworth, M. D. S. (1983). Patterns of infant–mother attachment as related to maternal care: Their early history and their contribution to continuity. In D. Magnusson & V. L. Allen (Eds.), *Human development: An interactional perspective.* Hillsdale, N.J.: Lawrence Erlbaum Associates.

Ajzen, J., & Fishbein, M. (1977). Attitude-behavior relations: A theoretical analysis and review of empirical research. *Psychological Bulletin, 84*, 888–918.

Allport, G. W. (1937). *Personality: A psychological interpretation.* New York: Holt, Rinehart, & Winston.

Allport, G. W. (1962). The general and the unique in psychological science. *Journal of Personality, 30*, 405–422.

Anastasi, A. (1958). Heredity, environment and the question "how?" *Psychological Review, 65*, 197–208.

Andersson, O., Dunér, A., & Magnusson, D. (1980). Social adjustment among early maturing girls. *Reports from the Department of Psychology, University of Stockholm*, Report No. 35.

Andersson, T., & Magnusson, D. (1986). Drinking habits and alcohol abuse from 15–25 years of age. Department of Psychology, University of Stockholm, manuscript, (In Swedish)

Andersson, T., Magnusson, D. & Dunér, A. (1983). Base Data-81: The life situation at early adulthood. *Reports from IDA*, Department of Psychology, University of Stockholm, No. 49. (In Swedish)

Angyal, A. (1941). *Foundations for a science of personality.* Cambridge, MA: Harvard University Press.

Appley, M. H., & Trumbull, R. (1967). *Psychological stress.* New York: Appleton-Century-Croft.

Aronfreed, J. (1968). *Conduct and conscience: The socialization of internalized control over behavior.* New York: Academic Press.

Arsenian, J., & Arsenian, J. M. (1948). Tough and easy cultures: A conceptual analysis. *Psychiatry, 11*, 377–385.

Åsberg, M., Mårtensson, B. & Wägner, A. (1987). Psychobiological aspects of suicidal behavior. In

D. Magnusson & A. Öhman (Eds.), *Psychopathology: An interactional perspective.* New York: Academic Press.

Åsberg, M., Schalling, D., Rydin, E., & Träskman-Bendz, L., (1983). Suicide and serotonin. In J. P. Soubrier & J. Vedrinne (Eds.), *Depression and suicide.* London: Pergamon Press.

Asmolov, A. G. (1984). The subject matter of psychology of personality. *Sovjet Psychology,* XXII, No. 4, 23–43.

Backenroth, G., & Magnusson, D. (1983). Basic results from the adult interview at age 27. *Reports from IDA.* Department of Psychology, University of Stockholm, No. 52. (In Swedish)

Backenroth, G., Magnusson, D., & Dunér, A. (1983). Planning and carrying through of the intensive study at age 27. *Reports from IDA,* Department of Psychology, University of Stockholm, No. 51 (In Swedish).

Backteman, G., & Magnusson, D. (1981). Longitudinal stability of personality characteristics. *Journal of Personality, 49,* 148–160.

Baldwin, A. L. (1969). A cognitive theory of socialization. In D. A. Goskin (Ed.), *Handbook of socialization. Theory and research.* Chicago: Rand McNally.

Baldwin, J. A. (1968). Psychiatric illness from birth to maturity: An epidemiological study. *Acta Psychiatrica Scandinavica, 44,* 313–333.

Baldwin, J. M. (1895). *Mental development in the child and the race: Method and processes.* New York: MacMillan.

Baldwin, J. M. (1897). *Social and ethical interpretations of mental development: A study in social psychology* (3rd ed.). New York: MacMillan.

Baltes, P. B. (1968). Longitudinal and cross-sectional sequencies in the study of age and generation effects. *Human Development, 11,* 145–171.

Baltes, P. B., Cornelius, S. W., & Nesselroade, J. R. (1979). Cohort effects in developmental psychology. In J. R. Nesselroade & P. B. Baltes (Eds.), *Longitudinal research in the study of behavior and development.* New York: Academic Press.

Baltes, P. B., & Nesselroade, J. R. (1973). The developmental analysis of individual differences on multiple measures. In J. R. Nesselroade & H. W. Reese (Eds.), *Life-span developmental psychology: Methodological issues.* New York: Academic Press.

Baltes, P. B., Reese, H. W., & Lipsitt, L. P. (1980). Life-span developmental psychology. In M. R. Rosenzweig & L. W. Porter (Eds.), *Annual Review of Psychology.* Palo Alto, CA: Annual Reviews.

Baltes, P. B., & Schaie, K. W. (1973). On life-span developmental research paradigms. In P. B. Baltes & K. W. Schaie (Eds.), *Life-span developmental psychology: Personality and socialization.* New York: Academic Press.

Bandura, A. (1978). The self system in reciprocal determinism. *American Psychologist,* 33, 344–358.

Bandura, A. (1980). Self-referent thought: A developmental analysis of self-efficacy. In J. H. Flavell & L. D. Ross (Eds.), *Cognitive social development: Frontiers and possible futures.* New York: Cambridge University Press.

Bandura, A. (1982). The psychology of chance encounters and life paths. *American Psychologist,* 37, 747–755.

Barchas, P. R. (Ed.) (1984). *Social hierarchies: Essays toward a sociophysiological perspective.* Westport, CT: Greenwood Press.

Barker, R. G. (1965). Exploration in ecological psychology. *American Psychologist,* 20, 1–14.

Barker, R. G., & Associates. (1978). *Habitats, environments and human behavior: Studies in ecological psychology and eco-behavioral science from the Midwest Psychological Field Station 1947–1972.* San Francisco: Jossey-Bass.

Bateson, P. P. G. (1978). How does behavior develop? In P. P. G. Bateson & P. H. Klopfer (Eds.), *Perspectives in ethology,* Vol. 3: *Social behavior.* New York: Plenum Press.

Bayley, N. (1949). Consistency and variability in the growth of intelligence from birth to eighteen years. *Journal of Genetic Psychology,* 75, 165–196.

Bayley, N. (1966). Developmental problems of the mentally retarded child. In J. Philips (Ed.), *Prevention and treatment of mental retardation*. New York: Basic Books.

Bell, R. Q. (1968). Reinterpretation of the direction of effects in studies of socialization. *Psychological Review, 75*, 81–95.

Bell, R. Q. (1971). Stimulus control of parent or caretaker by offspring. *Developmental Psychology, 4*, 63–72.

Bell, R. Q., Weller, G. M., & Waldrop, M. (1971). Newborn and preschooler: Organization of behavior and relations between periods. *Monographs of the Society for Research in Child Development, 36*, No. 142.

Berger, P., & Luckman, T. (1966). *The social construction of reality*. Garden City, N.J.: Doubleday.

Berglund, B. (1962). *Mental growth: A study of changes in test ability between the ages of nine and sixteen years*. Stockholm: Svenska Bokförlaget.

Bergman, L. R. (1971). Some univariate models in studying change. *Reports from the Psychological Laboratories, University of Stockholm, supplement*, No. 10.

Bergman, L. R. (1972a). Change as the dependent variable. *Reports from the Psychological Laboratories, University of Stockholm, supplement*, No. 14.

Bergman, L. R. (1972b). Linear transformations and the study of change. *Reports from the Psychological Laboratories, University of Stockholm*, No. 352.

Bergman, L. R. (1973). Parent's education and mean change in intelligence. *Scandinavian Journal of Psychology, 14*, 273–281.

Bergman, L. R. (1981). Is intellectual development more vulnerable in boys than in girls? *The Journal of Genetic Psychology, 138*, 175–181.

Bergman, L. R. (1985). A person approach to study development in an interactional perspective. In B. Törestad & L. Nystedt (Eds.), *Man–environment in interplay*. Stockholm: Natur och Kultur. (In Swedish).

Bergman, L. R., & El-Khouri, B. (1986). Some exact tests of single cell frequencies in two way contingency tables. *Reports from the Department of Psychology, University of Stockholm*, No. 645.

Bergman, L. R., Hanve, R., & Rapp, J. (1978). Why do some people refuse to participate in interview surveys? *Statistisk tidskrift, 5*, 341–356.

Bergman, L. R., & Magnusson, D. (1979). Overachievement and catecholamine excretion in an achievement-demanding situation. *Psychosomatic Medicine, 41*, 181–188.

Bergman, L. R., & Magnusson, D. (1983). The development of patterns of maladjustment. *Reports from IDA*. Department of Psychology, University of Stockholm, no. 50.

Bergman, L. R., & Magnusson, D. (1984a). Patterns of adjustment problems at age 10: An empirical and methodological study. *Reports from the Department of Psychology, University of Stockholm*, No. 615.

Bergman, L. R., & Magnusson, D. (1984b). Patterns of adjustment problems at age 13: An empirical and methodological study. *Reports from the Department of Psychology, University of Stockholm*, No. 620.

Bergman, L. R., & Magnusson, D. (1986). Type A behavior: A longitudinal study from childhood to adulthood. *Psychosomatic Medicine*, 134–142.

Bergman, L. R., & Magnusson, D. (1987). A person approach to the study of the development adjustment problems: An empirical example and some research strategical considerations. In D. Magnusson & A. Öhman (Eds.), *Psychopathology: An interactional perspective*. New York: Academic Press.

Bergman, L. R., & Magnusson, D. (in prep.) Multivariate development: The case of syndromes vs. single variables. Department of Psychology, University of Stockholm.

Berman, J. S., & Kenny, D. A. (1976). Correlational bias in observer ratings. *Journal of Personality and Social Psychology, 34*, 263–273.

Bertalanffy, L. von. (1968). *General system theory. Foundations, development, applications.* New York: Braziller.

Binet, A. (1909). *Les idées modernes sur les enfants* [Modern ideas about children]. Paris: Ernest Flamarion.

Bishop, Y. M. M., Feinberg, S. E., & Holland, P. L. (1975). *Discrete multivariate analysis: Theory and practice.* Cambridge: MIT Press.

Blalock, H. M. (1982). *Conceptualization and measurement in the social sciences.* Beverly Hills: Sage Publications.

Blancard, P., & Paynter, R. H. (1924). The problem child. *Mental Hygiene, 8*, 26–54.

Blane, H. (1979). Middle-aged alcoholics and young drinkers. In H. Blane & M. Chafez (Eds.), *Youth, alcohol and social policy.* New York: Plenum Press.

Block, J. (1971). *Lives through time.* Berkeley, CA: Bancroft Books.

Block, J. (1977). Correlational bias in observer ratings: Another perspective on the Berman–Kenny study. *Journal of Personality and Social Psychology, 35*, 873–880.

Block, J. (1982). Assimilation, accomodation and the dynamics of personality development. *Child Development*, 53, 281–295.

Block, J. H., & Block, J. (1980). The role of ego-control and ego-resiliency in the organization of behavior. In W. A. Collins (Ed.), *Minnesota Symposia on Child Psychology* (Vol. 13). Hillsdale, N.J.: Lawrence Erlbaum Associates.

Block, J., Weiss, D. S., & Thorne, A. (1979). How relevant is a semantic similarity interpretation of personality ratings? *Journal of Personality & Social Psychology*, 37, 1055–1074.

Bloom, B. S. (1964). *Stability and change in human characteristics.* New York: Wiley.

Bohman, M. (1978). Some genetic aspects of alcoholism and criminality. *Archives of General Psychiatry*, 35, 269–276.

Bolles, R. C. (1972). Reinforcement, expectancy and learning. *Psychological Review, 79*, 394–409.

Bowerman, W. R. (1978). Subjective competence: The structure, process and function of self-referent causal attribution. *Journal for the Theory of Social Behavior, 8*, 45–75.

Bowers, K. S. (1973). Situationism in psychology: An analysis and a critique. *Psychological Review, 80*, 307–336.

Bowlby, J. (1951). Maternal care and mental health. World Health Organization. *Monograph Series*, 2.

Boydstun, J. A., Ackerman, P. T., Stevens, D. A., Clemens, S. D., Peters, J. E., & Dykerman, R. A. (1968). Psychological and motor conditioning and generalization in children with minimal brain dysfunction. *Conditional Reflex*, 3, 81–104.

Bradley, R. H., & Caldwell, B. M. (1976). The relation of infants' home environments to mental test performance at fifty-four months: A follow-up study. *Child Development*, 47, 1172–1174.

Brandstädter, J. (1984). Personal and social control over development: Some implications of an action perspective in life-span developmental psychology. In P. B. Baltes & O. G. Brim (Eds.), *Life-span development and behavior.* (Vol. 6). New York: Academic Press.

Brim, O. G. Jr., & Kagan, J. (1980). *Constancy and change in human development.* Cambridge, Mass.: Harvard University Press.

Brim, O. G. Jr., & Ryff, C. D. (1980). On the properties of life events. In P. B. Baltes & O. G. Brim Jr. (Eds.), *Life-span development and behavior* (Vol. 3). New York: Academic Press.

Brocke, B. (1984). Diagnose, Ätiologie und Therapie des Hyperkinesen-Syndroms. *Praxis der Kinderpsychologie und Kinderpsychiatrie*, 33, 222–233.

Brooks-Gunn, J., & Petersen, A. C. (Eds.), (1983). *Girls at puberty.* New York: Plenum.

Bronfenbrenner, U. (1958). Socialization and social class through time and space. In E. E. Maccoby, T. M. Newcomb & E. L. Hartley (Eds.), *Readings in social psychology* (3rd Ed.). New York: Holt, Rinehart, & Winston.

Bronfenbrenner, U. (1977). Toward an experimental ecology of human development. *American Psychologist*, 32, 513–531.

Bronfenbrenner, U. (1979a). *The ecology of human development. Experiments by nature and design.* Cambridge, Mass.: Harvard University Press.

Bronfenbrenner, U. (1979b). Context of child rearing: Problems and prospects. *American Psychologist, 34,* 884–850.

Bronfenbrenner, U., & Crouter, A. C. (1983). The evolution of environmental models in developmental research. In W. Kassen (Ed.), *History, theories and methods.* (Vol. 1) of P. H. Mussen (Ed.), *Handbook of child psychology* (4th edition). New York: Wiley.

Brown, G. W., & Harris, T. (1978a). *Social origins of depression: A study of psychiatric disorder in women.* London: Tavistock Publications.

Brown, G. W., & Harris, T. (1978b). Social origins of depression: A reply. *Psychological Medicine, 8,* 577–588.

Brown, G. W., & Harris, T. (1980). Further comments on the vulnerability model. *British Journal of Psychiatry, 137,* 584–585.

Brunswik, E. (1952). *The conceptual framework of psychology.* Chicago: University of Chicago Press.

Buss, A. H., & Plomin, R. (1984). *Temperament: Early developing personality traits.* Hillsdale, N.J.: Lawrence Erlbaum Associates.

Cahalan, D., & Room, R. (1974). *Problem drinking among American men.* Rutger's center of alcohol studies. New Brunnswick.

Cairns, R. B. (1976). The ontogeny and phylogeny of social interactions. In M. Hahn & E. C. Simmel (Eds.), *Evolution of communicative behavior.* New York: Academic Press.

Cairns, R. B. (1977). Beyond social attachment. The dynamics of interactional development. In T. Alloway, P. Pliner & L. Krames (Eds.), *Attachment behavior.* New York: Plenum Press.

Cairns, R. B. (1979a). *Social development: The origins and plasticity of interchanges.* San Francisco: W. H. Freeman & Company.

Cairns, R. B. (1979b). Toward guidelines for interactional research. In R. B. Cairns (Ed.), *The analysis of social interactions: Methods, issues and illustrations.* Hillsdale, N.J.: Lawrence Erlbaum Associates.

Cairns, R. B. (1980). Developmental theory before Piaget. The remarkable contribution of James Mark Baldwin. *Contemporary Psychology, 25,* 438–440.

Cairns, R. B. (1983). The emergence of developmental psychology. In W. Kessen (Ed.), *History, theories and methods.* Vol. 1 of P. H. Mussen (Ed.), *Handbook of child psychology* (4th edition). New York: Wiley.

Cairns, R. B., & Cairns, B. D. (1985). The developmental-interactional view of social behavior: Four issues of adolescent aggression. In D. Olweus, J. Block & M. Radke-Yarrow (Eds.), *The development of antisocial and prosocial behavior.* New York: Academic Press.

Cairns, R. B., & Green, J. A. (1979). How to assess personality and social patterns: Observations or ratings. In R. B. Cairns (Ed.), *The analysis of social interactions: Methods, issues and illustrations.* Hillsdale, N.J.: Lawrence Erlbaum Associates.

Cairns, R. B., & Hood, K. E. (1983). Continuity in social development: A comparative perspective on individual difference prediction. In P. B. Baltes & O. G. Brim (Eds.), *Life-span development and behavior,* (Vol. 5). New York: Academic Press.

Cairns, R. B., & Nakelski, J. S. (1971). On fighting in mice: Ontogenetic and experimental determinants. *Journal of Comparative and Physiological Psychology, 71,* 354–364.

Cairns, R. B., & Ornstein, P. A. (1979). Developmental psychology. In E. Hearst (Ed.), *The first century of experimental psychology.* Hillsdale, N.J.: Lawrence Erlbaum Associates.

Cairns, R. B., & Valsiner, J. (1984). Child psychology. *Annual Review of Psychology, 35,* 553–577.

Carlsson, S. G., & Jern, S. (1982). Paradigms in psychosomatic research: A dialectic perspective. *Scandinavian Journal of Psychology,* Suppl. 1, 151–157.

Carroll, J. D., & Arabie, P. (1983). INDCLUS: An individual differences generalization of the ADCLUS model and the MAPCLUS algorithm. *Psychometrika, 48,* 157–170.

Cartwright, D. (1973). Determinants of scientific progress: The case of research in the risky shift. *American Psychologist*, 28, 222–231.

Cattell, R. B. (1963). Personality, role, mood, and situation perception: A unifying theory of modulators. *Psychological Review*, 70, 1–18.

Cattell, R. B. (1965). *The scientific analysis of personality*. Chicago: Aldine.

Cederblad, M. (1983). Epidemiological study of family functions and behavior disturbances in children age 3–15. *Barnombudsmannens årsskrift*, Stockholm: Rädda Barnen. (In Swedish)

Chein, I. (1954). The environment as a determinant of behavior. *Journal of Social Psychology*, 39, 115–127.

Cherry, N., & Rodgers, B. (1979). Using a longitudinal study to assess the quality of retrospective data. In L. Moss & H. Goldstein (Eds.), *The recall method in social surveys*. London: University of London, Inst. of Education, No. 9.

Clarke, A. D. B., & Clarke, A. M. (1984). Constancy and change in the growth of human characteristics. *Journal of Child Psychology and Psychiatry*, 25, 191–210.

Cofer, Ch. N. (1981). Introduction: Enduring issues and the nature of psychology. In R. A. Kasschau & Ch. N. Cofer (Eds.), *Psychology's second century: Enduring issues*. New York: Praeger.

Collett, J. (1963). Risk for recidivism in drunkenness. *Alkoholfrågan* 57:210. (In Swedish)

Cooley, C. H. (1902). *Human nature and the social order*. New York: Scribners.

Cox, A., Rutter, M., Yule, B., & Quinton, D. (1977). Bias resulting from missing information: Some epidemiological findings. *British Journal of Preventive Social Medicine*, 31, 131–136.

Cronbach, L. J. (1975). Beyond the two disciplines of scientific psychology. *American Psychologist*, 30, 116–127.

D'Andrade, R. G. (1974). Memory and assessment of behavior. In H. M. Blalock, Jr. (Ed.), *Measurement in the social sciences*. Chicago: Aldine.

Davies, B. L. (1977). Attitudes toward school among early and late-maturing adolescent girls. *Journal of Genetic Psychology*, 131, 261–266.

Davis, A. J., & Hathaway, B. K. (1982). Reciprocity in parent-child verbal interactions. *Journal of Genetic Psychology*, 140. 169–183.

Depue, R. A., Monroe, S. M. & Shackman, S. L. (1979). The psychobiology of human disease: Implications for conceptualizing the depressive disorders. In R. A. Depue (Ed.), *The psychobiology of depressive disorders*. New York: Academic Press.

Dewey, J. (1896). The reflex arc concept in psychology. *Psychological Review*, 3, 357–370.

Diagnostic and statistical manual of mental disorders (1980). 3rd ed. Washington: American Psychiatric Association.

Diener, E., & Larsen, R. J. (1984). Temporal stability and cross-situational consistency of affective, behavioral and cognitive responses. *Journal of Personality and Social Psychology*, 47, 871–883.

Dohrenwend, B. S., & Dohrenwend, B. P. (Eds.) (1974). *Stressful life events: Their nature and effects*. New York: Wiley.

Donovan, J., Jessor, R., & Jessor, L. (1983). Problem drinking in adolescence and young adulthood: A follow-up study. *Journal of Studies on Alcohol*, Vol. 44, Nr 1.

Douglas, V. J. (in press). Attention deficit disorder: Are we any further ahead? *Canadian Journal of Behavioral Psychology*.

Dunér A. *(Ed.)* (1978a). *Research into personal development: educational and vocational choice*. Amsterdam and Lisse: Swets & Zeitlinger.

Dunér, A. (1978b). Problems and designs in research on educational and vocational career. In A. Dunér (Ed.). *Research into personal development: Educational and vocational choice*. Amsterdam and Lisse: Swets & Zeitler.

Dunér, A., & Magnusson, D. (1979). Achievement, social adjustment and home background. *Reports from the Department of Psychology, University of Stockholm*, No. 547.

Dunn, J. (1981). Maturation and early social development. In K. J. Connolly & H. R. Prechtl (Eds.), *Maturation and development: Biological and psychological perspectives*. London: Heinemann Medical Books.

Eichorn, D. H. (1975). Asynchronizations in adolescent development. In S. E. Dragastin & G. H. Elder (Eds.), *Adolescence in the life cycle: Psychological change and social context*. New York: Wiley.

Ekehammar, B. (1974). Interactionism in personality from a historical perspective. *Psychological Bulletin, 81*, 1026–1048.

Ekehammar, B. (1977a). Test of a psychological cost - benefit model for career choice. *Journal of Vocational Behavior, 10*, 245–260.

Ekehammar, B. (1977b). Intelligence and social background as related to psychological cost - benefit in career choice. *Psychological Reports, 40*, 963–970.

Ekehammar, B. (1978a). Toward a psychological cost - benefit model for educational and vocational choice. *Scandinavian Journal of Psychology, 19*, 15–27.

Ekehammar, B. (1978b). Psychological cost - benefit as an intervening construct in career choice models. *Journal of Vocational Behavior, 12*, 279–289.

Ekehammar, B., & Magnusson, D. (1973). A method to study stressful situations. *Journal of Personality and Social Psychology, 27*, 176–179.

Ekehammar, B., Schalling, D., & Magnusson, D. (1975). Dimensions of stressful situations: A comparison between a response analytical and a stimulus analytical approach. *Multivariate Behavioral Research, 10*, 155–164.

Eklund, L., & Nylander, I. (1965). Risk for recidivism in drunkenness among Stockholm boys. *Socialmedicinsk Tidskrift, 42*, 201–205. (In Swedish)

Elder, G. H. (1979). Historical change in life patterns and personality. In P. B. Baltes (Ed.), *Life-span development and behavior* (Vol. 2). New York: Academic Press.

Emmerich, W. (1964). Continuity and stability in early social development. *Child Development, 35*, 311–332.

Emmerich, W. (1968). Personality development and concepts of structure. *Child Development, 39*, 671–690.

Endler, N. S. (1975). A person-situation interaction model of anxiety. In C. D. Spielberger & J. G. Sarason (Eds.), *Stress and anxiety* (Vol. 1). Washington D.C.: Hemisphere.

Endler, N. S. (1983). A personality model, but not yet a theory. In M. M. Page (Ed.), *Nebraska Symposium on Motivation 1982: Personality, current theory and research*. Lincoln, Nebraska: University of Nebraska Press.

Endler, N. S. (1984). Interactionism. In N. S. Endler & J. McV. Hunt (Eds.), *Personality and the behavioral disorders*. (Vol 1). New York: Wiley.

Endler, N. S., & Magnusson, D. (1976a). Toward an interactional psychology of personality. *Psychological Bulletin, 83*, 956–979.

Endler, N. S., & Magnusson, D. (1976b). Personality and person by situation interactions. In N. S. Endler & D. Magnusson (Eds.), *Interactional psychology and personality*. Washington D.C.: Hemisphere.

Epstein, S. (1979). The stability of behavior: On predicting most of the people much of the time. *Journal of Personality and Social Psychology, 37*, 1097–1126.

Epstein, S. (1980). The self-concept: A review and the proposal of an integrated theory of personality. In E. Staub (Ed.), *Personality: Basic issues and current research*. Englewood Cliffs, N.J.: Prentice Hall.

Everitt, B. (1977). *Cluster analysis*, London: Heinemann.

Eysenck, H. J. (1967). *The biological basis of personality*. Springfield, JU.: Thomas.

Eysenck, H. J. (1983). Is there a paradigm in personality research? *Journal of Research in Personality, 17*, 369–397.

Fahrenberg, J. (1984). Psychophysical individuality: A pattern analytic approach to personality research and psychosomatic medicine. Psychological Institute, University of Freiburg, Report No. 16.

Farrington, D. P. (1979). Longitudinal research on crime and delinquency. In N. Morris & M.

Tonry (Eds.), *Criminal justice: Annual review of research*, Vol. 1, 284–348. Chicago: University of Chicago Press.

Farrington, D. (1985). Delinquency prevention in the 1980s. *Journal of Adolescence, 8*, 3–16.

Faust, M. S. (1960). Developmental maturity as a determinant of prestige in adolescent girls. *Child Development, 31*, 173–184.

Featherman, D., & Lerner, R. M. (1985). Ontogenesis and sociogenesis: Problematics for theory and research about development and socialization across the life span. *Scholarly Report Series: Center for the Study of Child and Adolescent Development*. The Pennsylvania State University. No. 6.

Feldhusen, J. F., Thurston, J. R. & Benning, J. J. (1973). A longitudinal study of delinquency and other aspects of children's behavior. *International Journal of Criminology and Penology, 1*, 341–351.

Fillmore, K. (1974). Drinking and problem drinking in early adulthood and middle age: An exploratory 20-years follow-up study. *Quarterly Journal of Studies on Alcohol*, 35, 819–840.

Fishbein, S. (1979). *Heredity-environment influences on growth and development during adolescence*. Lund CWK/Gleerups, Studies in education and psychology, No. 4.

Fishman, D. B., & Neigher, W. D. (1982). American psychology in the eighties: Who will buy? *American Psychologist*, 37, 533–546.

Fishman, D. B., & Peterson, D. R. (in press). On getting the right information and getting the information right. In D. R. Peterson & D. B. Fishman (Eds.), *Assessment for decision*. Rutgers University: Rutgers University Press.

Flavell, J. H. (1971). Age related properties of cognitive development. *Cognitive psychology*, 2, 421–453.

Flavell, J. H. (1982). Structures, stages and sequences in cognitive development. In W. A. Collins (Ed.), *The concept of development. The Minnesota Symposia on Child Psychology*. Vol. 15. Hillsdale, N.J.: Lawrence Erlbaum Associates.

Frisk, M., Tenhunen, T., Widholm, O. & Hortling, H. (1966). Physical problems in adolescents showing advanced or delayed physical maturation. *Adolescence, 1*, 126–140.

Galton, F. (1865). Hereditary talent and character. London: *MacMillan's Magazine, 12*, 157–166.

Galton, F. (1869). *Hereditary genius: An inquiry into its laws and consequences*, London: Mac Millan.

Gardner, H. (1983). *Frames of mind*. New York: Basic Books.

Garmezy, N. (1976). *Vulnerable and invulnerable children: Theory, research and intervention*. Master lecture on developmental psychology. Washington; D.C.: American Psychological Association, No. 1337.

Garmezy, N. (1981). Children under stress: Perspectives in antecedents and correlates of vulnerability and resistance to psychopathology. In A. J. Robin, J. Arowoff, A. M. Barclay & R. A. Zucker (Eds.), *Further explorations in personality*. New York: Wiley.

Garmezy, N. (1983). Stressors of childhood. In N. Garmezy & M. Rutter (Eds.), *Stress, coping and development in children*. New York: McGraw-Hill.

Garmezy, N., & Nuechterlein, K. (1972). Invulnerable children: The fact and fiction of competence and disadvantage. *American Journal of Ortopsychiatry*, 42, 328–329. (Abstract)

Garmezy, N., & Tellegen, A. (1984). Studies of stress-resistant children: Methods, variables and preliminary findings. In F. Morrison, C. Lord & D. Keating (Eds.), *Applied developmental psychology*, Vol. 1. New York: Academic Press.

Garn, S. H. (1980). Continuities and change in maturational timing. In O. G. Brim & J. Kagan (Eds.), *Constancy and change in human development*. Cambridge, MA.: Harvard University Press.

Gergen, K. J. (1973). Toward generative theory. *Journal of Personality and Social Psychology*, 26, 309–320.

Gesell, A. L. (1928). *Infancy and human growth*. New York: Mac Millan.

Gesell, A. L., & Thompson, H. (1934). *Infant behavior: Its genesis and growth.* New York: McGraw-Hill.

Goffman, E. (1964). The neglected situation. *American Anthropologist, 66,* 133–136.

Goldberg, L. R. (1978). Differential attribution of trait-descriptive terms to oneself as compared to well-liked neutral, and disliked others. A psychometric analysis. *Journal of Personality and Social Psychology, 36,* 1012–1029.

Goldstein, H. (1979). *The design and analysis of longitudinal studies.* New York: Academic Press.

Goodfield, J. (1974). Changing strategies: A comparison of reductionist attitudes in biological and medical research in the nineteenth and twentieth centuries. In F. J. Ayala & T. Dobzhansky (Eds.), *Studies in the philosophy of biology.* Berkeley: University of California.

Goodwin, D. W., Schulzinger, F., Moller, N., Hermansen, L., Winokur, G., & Guze, S. B. (1974). Drinking problems in adopted and non-adopted sons of alcoholics. *Archives of General Psychiatry, 31,* 164–169.

Gottlieb, G. (1976a). Conceptions of prenatal development: Behavioral embryology. *Psychological Review, 83,* 215–234.

Gottlieb, G. (1976b). The roles of experience in the development of behavior and the nervous system. In G. Gottlieb (Ed.), *Neural and behavioral specificity.* New York: Academic Press.

Gottlieb, G. (1981). Roles of early experience in species-specific perceptual development. In R. N. Aslin, J. R. Alberts, & M. R. Petersen (Eds.), *Development of perception (Vol. 1).* New York: Academic Press.

Gottlieb, G. (1983). The psychobiological approach to developmental issues. In P. H. Mussen (Ed.), *Handbook of child psychology.* Vol: II. (Volume editors M. M. Haith and J. J. Campos). New York: Wiley.

Gould, S. D., & Vrba, E. (1982). Exaptation: A missing term in the science of form. *Paleobiology, 8,* 4–15.

Graham, P., & Rutter, M. (1976) Adolescent disorder. In M. Rutter & L. Hersov (Eds.) *Child psychiatry: Modern approaches.* Oxford: Blackwell Scientific Publications.

Guilford, J. P. (1959). *Personality.* McGraw Hill: Book Company Inc.

Gustafsson, T., & Toneby, M. J. (1971). How genes control morphogenesis: The role of serotonin and acetylcholine in morphogenesis. *American Scientist, 59,* 452–462.

Hamburg, D. A., Elliott, G. R., & Parron, D. L. (1982). *Health and behavior: Frontiers of research in the behavioral sciences.* Washington, D.C.: National Academia Press.

Harris, C. W. (Ed.), (1963). *Problems in measuring change.* Madison WI: University of Wisconsin Press.

Harter, S. (1983). Developmental perspectives on the self system. In E. M. Hetherington (Ed.), *Socialization, personality and social development.* Vol. 4 of P. H. Mussen (Ed.), *Handbook of Child Psychology.* New York: Wiley, pp. 275–385.

Hartup, W. W. (1978). Perspectives on child and family interaction: Past, present and future. In R. M. Lerner & G. B. Spanier (Eds.), *Child inferences on marital and family interaction.* New York: Academic Press.

Hartup, W. W. (1979). Peer relations and the growth of social competence. In M. W. Kent & J. E. Rolf (Eds.), *Primary prevention of psychopathology. Social competence in children* (Vol. 3). Hanover, NH: University Press of New England.

Havighurst, R. J., Bowman, P. H., Liddle, G. P., Matthews, C. V. & Pierce, J. V. (1962). *Growing up in River City.* New York: Wiley.

Hebb, D. D. (1958). The socialization of the child. In E. E. Maccoby, T. M. Newcomb & E. L. Hartley (Eds.), *Readings in social psychology.* New York: Holt, Rinehart and Winston.

Heckhausen, H. (1983). Concern with one's competence. Developmental shifts in person-environment interaction. In D. Magnusson & V. L. Allen (Eds.), *Human development: An interactional perspective.* Hillsdale, N.J.: Lawrence Erlbaum Associates.

Heffler, B., & Magnusson, D. (1979). The generality of behavioral data IV: Cross-situational invariance of objectively measured behavior. *Perceptual and Motor Skills, 48,* 471–477.

Hetherington, E. M., Cox, M., & Cox, R. (1979). Play and social interaction in children following divorce. *Journal of Social Issues*, 35, 26–49.
Hibell, B. (1977) *On the development of youth alcohol habits from 1947 to 1976*. Lund: Doxa. (In Swedish)
Hinde, R. A., & Bateson, P. (1984) Discontinuities versus continuities in behavioral development and the neglect of process. *International Journal of Behavioral Development*, 7, 129–143.
Hofer, M. A. (1981). *The roots of human behavior. An introduction to the psychobiology of early development*. San Francisco: W. H. Freeman and Company.
Hofer, M. A. (1982). Some thoughts on "the transduction of experience" from a developmental perspective. *Psychosomatic Medicine*, 44, 19–28.
Hubel, D. H., & Wiesel, T. N. (1970). The period of susceptibility to the physiological effects of unilateral eye closure in kittens. *Journal of Physiology*, 206, 419–436.
Hultsch, D. F., & Plemons, J. K. (1979). Life events and life-span development. In P. B. Baltes & O. G. Brian (Eds.), *Life-span development and behavior*. Vol. 2. New York: Academic Press.
Hunt, J. McV. (1961). *Intelligence and experience*. New York: The Ronald Press Company.
Hunt, J. McV. (1962). Motivation inherent in information processing and action. In O. J. Harvey (Ed.), *Cognitive factors in motivation and social organization*. New York: Ronald Press Company.
Hunt, J. McV. (1979). Psychological development: Early experience. *Annual Review of Psychology*, 30, 103–143.
Hunt, J. McV. (1981). The role of situations in early psychological development. In D. Magnusson (Ed.), *Toward a psychology of situations*. Hillsdale, N.J.: Lawrence Erlbaum Associates.
Husén, T. (1981). Why conduct longitudinal research. In V. Saloheimo (Ed.), *Publications of the University of Joensuu, Series* A, Vol. 20.
Hutt, C. (1972). Sex differences in human development. *Human Development*, 15, 153–170.
James, L. R. (1982). Aggregation bias in estimates of perceptual agreement. *Journal of Applied Psychology*, 67, 219–229.
James, W. (1890). *The principles of psychology*. New York: Holt.
Janson, C.-G. (1981). The longitudinal approach. In F. Schulzinger, S. A. Mednick & J. Knop (Eds.), *Longitudinal research: Methods and uses in behavioral science*. Boston: Nijhoff.
Jemmott III, J. B. & Locke, S. E. (1984). Psychosocial factors, immunologic mediation and human susceptibility to infection diseases: How much do we know? *Psychological Bulletin*, 95, 78–108.
Jenkins, C. D. (1985). New horizons for psychosomatic medicine. *Psychosomatic Medicine*, 47, 3–25.
Jessor, R. (1956). Phenomenological personality theories and the data language of psychology. *Psychological Review*, 63, 173–180.
Jessor, R. (1981). The perceived environment in psychological theory and research. In D. Magnusson (Ed.), *Toward a psychology of situations: An interactional perspective*. Hillsdale, N.J.: Lawrence Erlbaum Associates.
Jessor, R. (1984). Adolescent development and behavioral health. In J. D. Matarazzo, S. M. Weiss, J. A. Herd, N. E. Miller & S. M. Weiss (Eds.), *Behavioral health: A handbook of health enhancement and disease prevention*. New York: Wiley.
Jessor, R., & Jessor, S. L. (1977). *Problem behavior and psychosocial development: A longitudinal study of youth*. New York: Academic Press.
Johansson, G., Frankenhauser, M. & Magnusson, D. (1973). Catecholamine output in school children as related to performance and adjustment. *Scandinavian Journal of Psychology*, 14, 20–28.
Jonsson, G., & Kälvesten, A. L. (1964). 222 *Stockholm boys*. Uppsala: Almqvist & Wiksell. (In Swedish)
Jöreskog, K. G., & Sörbom, D. (1981). LISREL V. Analysis of linear structural relationships by maximum likelihood and least square methods. *Research reports from the department of Statistics, University of Uppsala*, Sweden, No. 81–8.

Kagan, J. (1967). On the need for relativism. *American Psychologist, 22*, 131–142.
Kagan, J. (1971). *Change and continuity in infancy.* New York: Wiley.
Kagan, J. (1978). Continuity and change in human development. In P. P. G. Bateson & P. H. Klopfer (Eds.). *Perspectives in ethology. Vol. 3. Social behavior.* New York: Plenum Press.
Kagan, J. (1983). Stress and coping in early development. In N. Garmezy & M. Rutter (Eds.), *Stress, coping and development in children.* New York: Mc Graw-Hill.
Kagan, J., & Moss, H. A. (1962). *Birth to maturity: A study in psychological development.* New York: Wiley.
Kalverboer, A. F. & Hopkins, B. (1983). General introduction: A biopsychological approach to the study of human behavior. *Journal of Child Psychology and Psychiatry, 24*, 9–10.
Kamin, L. J. (1978). Sex differences in susceptibility of IQ to environmental influence. *Child Development, 49*, 517–518.
Kamin, L. J. (1985). Genes and behavior: The missing link, *Psychology Today*, Oct. 1985, 76–78.
Kantor, J. R. (1924). *Principles of psychology. Vol. 1.* Bloomington: Principia Press.
Kantor, J. R. (1926). *Principles of Psychology. Vol. 2.* Bloomington: Principia Press.
Kellam, S. G., Brown, C. H., Rubin, B. R., & Ensminger, M. E. (1983). Paths leading to teenage psychiatric symptoms and substance use: Developmental epidemiological studies in Woodlawn. In S. B. Guze, F. J. Earls & J. E. Barrett (Eds.), *Childhood psychology and development.* New York: Raven Press.
Kelly, G. A. (1955). *The psychology of personal constructs.* New York: Norton.
Kenny, D. A., & Judd, C. (1984). Estimating the nonlinear and interactive effects of latent variables. *Psychological Bulletin, 96*, 201–210.
Kessen, W. (1979). The American child and other culture inventions. *American Psychologist, 34, 10*, 815–820.
Kirkegaard-Sorensen, L., & Mednick, S. A. (1977). A prospective study of predictors of criminality. In S. A. Mednick & K. O. Christiansen (Eds.), *Biosocial bases of criminal behavior.* New York: Gardner Press.
Kitcher, P. (1985). *Vaulting ambition: Sociobiology and the quest for human nature.* MIT Press.
Kløve, H., & Hole, K. (1979). The hyperkinetic syndrome. Criteria for diagnosis. In R. L. Trites (Ed.), *Hyperactivity in children. Etiology, measurement and treatment implications.* Baltimore: University Park Press.
von Knorring, A-L., Andersson, O., & Magnusson, D. (1987). Psychiatric care and course of psychiatric disorders from childhood to early adulthood in a representative sample. *Journal of Child Psychology and Psychiatry.* 28, 329–341.
Koch, S. (1959). Epilogue. In S. Koch (Ed.), *Psychology: A study of a science.* Vol. 3. New York: McGraw-Hill.
Koch, S. (1961). Psychological science versus the science-humanism antimony: Intimations of a significant science of man. *American Psychologist, 16*, 629–639.
Koch, S. (1981). The nature and limits of psychological knowledge: Lessons of a century qua "science." *American Psychologist.* 36, 257–269.
Koffka, K. (1935). *Principles of Gestalt Psychology.* New York: Harcourt.
Kohlberg, L. (1969). Stage and sequence: The cognitive-developmental approach to socialization. In D. A. Goolin (Ed.), *Handbook of socialization theory and research.* Chicago: Rand McNally.
Kohnen, R., & Lienert, G. A. (1987). Interactional research in human development approached by interactive configural frequency analysis. In D. Magnusson & A. Öhman (Eds.), *Psychopathology: An interactional perspective.* New York: Academic Press.
Krauth, J., & Lienert, G. A. (1973). *Die Konfigurationsfrequenzanalyse und ihre Anwendung in Psychologie und Medicin.* Munchen: Verlag Karl Alber.
Krauth, J., & Lienert, G. A. (1982). Fundamentals and modifications of configural frequency analysis (CFA). *Interdisciplinaria,* 3, issue 1.
Kuo, Z-Y. (1967). *The dynamics of behavior development: An epigenetic view.* New York: Random House.

Lagerspetz, K. M. J., & Lagerspetz, K. Y. H. (1971). Changes in the aggressiveness of mice resulting from selective breeding, learning and social isolation. *Scandinavian Journal of Psychology, 12*, 241–248.

Lagerström, M., Nyström, B., Bremme, K., Magnusson, D., & Eneroth, P. (1985). Basic data of delivery records. Organization of the study. *Reports from IDA*. Department of Psychology, University of Stockholm, Report No. 62.

Laing, R., & Esterson, A. (1964). *Sanity, madness and the family* Vol. 1. London: Tavistock.

Lambert, W. W., Johansson, G., Frankenhaeuser, M., & Klackenberg-Larsson, I. (1969). Cathecholamine excretion in young children and their parents as related to behavior. *Scandinavian Journal of Psychology, 10*, 306–318.

Lamiell, J. T. (1982). The case for an idiothetic psychology of personality: Conceptual and empirical foundation. In B. Marer (Ed.), *Progress in experimental personality research*, Vol. 11.

Lander, J. H., Wallace, J. A., & Krebs, H. (1981). Roles for serotonin in neuroembryogenesis. In B. Haber, S. Gaby, M. R. Issidorides, & S. G. A. Alivivatos (Eds.), *Serotonin: Current aspects of neurochemistry and function*. Advances in experimental medicine and biology. New York: Plenum.

Langer, J. (1969). *Theories of development*. New York: Holt, Rinehart & Winston, Inc.

Lavik, N. J. (1976). *Ungdomsmentale helse*. Oslo: Universitetsforlaget.

Lazarsfeld, P. F., & Henry, N. W. (1968). *Latent structure analysis*. New York: Houghton Mifflin Company.

Lazarus, R. S., & Launier, R. (1978). Stress-related transactions between person and environment. In L. A. Pervin & M. Lewis (Eds.), *Perspectives in interactional psychology*. New York: Plenum.

Leontief, W. (1982). Academic economics. *Science* 217(1), 104–105.

Lerner, R. M. (1976). *Concepts and theories of human development*. Reading, Mass.: Addison-Wesley.

Lerner, R. M. (1978). Nature, nurture and dynamic interactionism. *Human Development, 21*, 1–20.

Lerner, R. M. (1984). *On the nature of human plasticity*. Cambridge: Cambridge University Press.

Lerner, R. M., & Busch-Rossnagel, N. A. (1981). Individuals as producers of their development: Conceptual and empirical bases. In R. M. Lerner & N. A. Busch-Rossnagel (Eds.), *Individuals as producers of their development: A life-span perspective*. New York: Academic Press.

Lerner, R. M., & Kauffman, M. B. (1985). The concept of development in contextualism. *Developmental Review*, 5, 309–333.

Lerner, J. V., & Lerner, R. M. (1983). Temperament and adaptations across life: Theoretical and empirical issues. In P. B. Baltes & O. G. Brim, Jr. (Eds.), *Life-span development and behavior*. (Vol. 5). New York: Academic Press.

Lerner, R. M., Palermo, M., Spiro III, A., & Nesselroade, J. R. (1982). Assessing the dimension of temperamental individuality across the life span: The dimensions of temperament survey (DOTS). *Child Development*, 53, 1–19.

Lerner, R. M., Skinner, E. A., & Sorrell, G. T. (1980). Methodological implications of contextual/dialectic theories of development. *Human Development*, 23, 225–235.

Levine, S. (1982). Comparative and psychobiological perspectives on development. In W. A. Collins (Ed.), *The concept of development, The Minnesota Symposia on Child Psychology*. Vol. 15. Hillsdale, N.J.: Lawrence Erlbaum Associates.

Lewin, K. (1931). Environmental forces. In C. Murchison (Ed.), *A handbook of child psychology*. Worcester, MA: Clark University Press.

Lewin, K. (1935). *A dynamic theory of personality*. New York: McGraw Hill.

Lewis, M., & Brooks-Gunn, J. (1979). *Social cognition and the acquisition of self*. New York: Plenum.

Lidberg, L., Levander, S. E., Schalling, D., & Lidberg, Y. (1978). Urinary catecholamines, stress and psychopathy: A study of arrested men awaiting trial. *Psychosomatic Medicine*, 40, 116–125.

Lienert, G. A. (1969). Die Konfigurationsfrequenzanalyse als Klassifikationsmittel in der Klinischen Psychologie In M. Irle (Ed.), *Bericht uber den 26 Kongress der Deutschen Gesellschaft für Psychologie*, Tübigen, Göttingen: Hogrefe.

Lienert, G. A., & Bergman, L. R. (in press). Longisectional interaction analysis in clinical and experimental psychopathology. *Neuropsychobiology*.

Lienert, G. A., & zur Oeveste, H. (1985). CFA as a statistical tool for developmental research. *Educational & Psychological Measurement*, 301–307.

Lindgren, G. (1976). Height, weight, and menarche in Swedish urban school children in relation to socio-economic and regional factors. *Annals of Human Biology*, 3, 501–528.

Lipsitt, L. P. (1983). Stress in infancy: Toward understanding the origins of coping behavior. In N. Garmezy & M. Rutter (Eds.), *Stress, coping and development in children*. New York: Mc Graw Hill.

Livson, N., & Peskin, H. (1980). Perspectives on adolescence from longitudinal research. In J. Adelson (Ed.), *Handbook of Adolescent Psychology*. New York: Wiley.

Ljung, B.-O. (1965). *The adolescent spurt in mental growth*. Stockholm: Almqvist & Wiksell.

Loeber, R., & Dishion, T. (1983). Early prediction of male delinquency: A review. *Psychological Bulletin*, *94*, 68–99.

Loehlin, J. C. (1982). Are personality traits differentially heritable? *Behavior Genetics*, *12*, 417–427.

Loevinger, J. (1966). Models and measures of developmental variation. In J. Brozek (Ed.), Biology of human variation. *Annals of the New York Academy of Sciences*, *134*, 585–590.

Loney, J., Langhorne, J. E., & Paternite, C. E. (1978). An empirical basis for subgrouping the hyperkinetic/minimal brain disfunction syndrome. *Journal of Abnormal Psychology*, 87, 431–441.

Loper, R., Kammier, M. L., & Hoffman, H. (1973). MMPI characterstics of college freshmen who later became alcoholics. *Journal of Abnormal Psychology*, 82, 159–162.

Lykken, D. T. (1982). Fearlessness: It's carefree charm and deadly risks. *Psychology Today*, *16*, 20–28.

Maccoby, E. E., & Jacklin, C. N. (1983). The "person" characteristics of children and the family as environment. In D. Magnusson & V. L. Allen (Eds.), *Human development: An interactional perspective*. Hillsdale, N.J.: Lawrence Erlbaum Associates.

McCall, R. B. (1977). Challenges to a science of developmental psychology. *Child Development*, *48*, 333–344.

McCall, R. B. (1981). Nature, nurture and the two realms of development: A proposed integration with respect to mental development. *Child Development*, 52, 1–12.

McCall, R. B., Appelbaum, M. I., & Hagerty, P. S. (1973). Developmental changes in mental performance. *Monographs of the Society for Research in Child Development*, 38 (3, whole No. 150).

McClelland, D. C. (1955). *Studies in motivation*. New York: Appleton-Century-Croft.

McClintock, E. (1983). Interaction. In H. H. Kelley, E. Berscheid, A. Christensen, J. H. Harvey, T. L. Huston, G. Levinger, E. M. McClintock, L. A. Peplau, & D. R. Peterson (Eds.), *Close relationships*. New York: W. H. Freeman and Company.

McCord, J. A. (1983). A longitudinal study of aggression and antisocial behavior. In K. T. Van Dusen & S. A. Mednick (Eds.), *Prospective studies of crime and delinquency*. Boston: Kluwer-Nijhoff.

McGee, R., Silva, P. A., & Williams, S. (1984). Behavior problems in a population of seven-year-old children: Prevalence, stability and types of disorder—a research report. *Journal of Child Psychology and Psychiatry*, 25, 251–260.

Mc Guire, M. T., Raleigh, H. J., & Brammer, G. L. (1982). Sociopharmacology. *Annual Review of Pharmacology and Toxicology*, 22, 643–661.

Mc Guire, M. T., Raleigh, M. J., & Johnson, C. (1983a). Social dominance in adult vervet monkeys: Behavior-biochemical relationships. *Social Science Information*, 22, 311–328.

Mc Guire, M. T., Raleigh, M. J., & Johnson, C. (1983b). Social dominance in adult vervet monkeys: General considerations. *Social Science Information*, 22, 89–123.

Mc Gulloc, J. W., Henderson, A. S., & Philip, A. E. (1966). Psychiatric illness in Edinburgh teenagers. *Scottish Medical Journal*, *11*, 277–281.

Mc Nemar, Q. (1960). At random: Sense and nonsense. *American Psychologist*, *15*, 295–300.

Magnusson, D. (1960). Self-evaluation and school environment. In I. Johannesson & D. Magnusson, *Social- and personality psychological factors in relation to school differentiation*. SOU:42. Stockholm: Allmänna Förlaget. (In Swedish)

Magnusson, D. (1963). *Maladjustment and structure of intelligence*. Stockholm: Skandinaviska Testförlaget (In Swedish).

Magnusson, D. (1967). *Test theory*. Reading, Mass.: Addison-Wesley.

Magnusson, D. (1971). An analysis of situational dimensions. *Perceptual and Motor Skills*, 32, 851–867.

Magnusson, D. (1974). The individual in the situation: Some studies on individuals' perception of situations. *Studia Psychologica*, *XVI*, 2, 124–132.

Magnusson, D. (1976). The person and the situation in an interactional model of behavior. *Scandinavian Journal of Psychology*, *17*, 253–271.

Magnusson, D. (1980). Personality in an interactional paradigm of research. *Zeitschrift für Differentielle und Diagnostische Psychologie*, *1*, 17–34.

Magnusson, D. (Ed.), (1981a). *Toward a psychology of situations: An interactional perspective*. Hillsdale, NJ: Lawrence Erlbaum Associates.

Magnusson, D. (1981b). Problems in environmental analyses-an introduction In D. Magnusson (Ed.), *Toward a psychology of situations: An interactional perspective*. Hillsdale, N.J.: Lawrence Erlbaum Associates.

Magnusson, D. (1984a). On the situational context. In K. M. J. Lagerspetz & P. Niemi (Eds.), *Psychology in the 1990's*. Amsterdam: Elsevier Science Publishers.

Magnusson, D. (1984b). The situation in an interactional paradigm of personality research. In V. Sarris & A. Parducci (Eds.), *Perspectives in psychological experimentation: Toward the year 2000*. Hillsdale, N.J.: Lawrence Erlbaum Associates.

Magnusson, D. (1985a). Man in society: A psychological and biological being in interaction with his environment. *Reports from the Department of Psychology, University of Stockholm*, No 634.

Magnusson, D. (1985b). Implications of an interactional paradigm for research on human development. *International Journal of Behavior Development*, 8, 115–137.

Magnusson, D. (1985c). Early conduct and biological factors in the developmental background of adult delinquency. *The British Psychological Society, Newsletter*, 13, 1, 4–17.

Magnusson, D. (1987). Adult delinquency in the light of conduct and physiology at an early age. In D. Magnusson & A. Öhman (Eds.), *Psychopathology: An interactional perspective*. New York: Academic Press.

Magnusson, D., & Allen, V. (Eds.) (1983a). *Human development: An interactional perspective*. New York: Academic Press.

Magnusson, D., & Allen, V. (1983b). An interactional perspective for human development. In D. Magnusson & V. Allen (Eds.), *Human development: An interactional perspective*. New York: Academic Press.

Magnusson, D., & Backteman, G. (1978). Longitudinal stability of person characteristics: Intelligence and creativity. *Applied Psychological Measurement*, 2, 481–490.

Magnusson, D., & Beckne, R. (1967). Summary of earlier research. Reports from IDA. Department of Psychology, University of Stockholm, No. 2 (In Swedish).

Magnusson, D., & Bergman, L. R. (1984). On the study of the development of adjustment problems. In L. Pulkkinen & P. Lyytinen (Eds.), *Human action and personality essays in honour of Martti Takala*, Jyväskylä Studies in Education, Psychology and Social Research. Jyväskylä, Finland: University of Jyväskylä.

Magnusson, D., & Bergman, L. R. (in press). Longitudinal studies: Individual and variable based approaches to research on early risk factors. In M. Rutter (Ed.), *Risk and protective factors in psychosocial development*. (Preliminary title of a book to arise out of the Minster Lovell Workshop on Early Risk Factors, January 28–31, 1987, to be published by Cambridge University Press.)

Magnusson, D., & Dunér, A. (1967). Methods and models. *Reports from IDA*. Department of Psychology, University of Stockholm, No. 3 (In Swedish).

Magnusson, D., & Dunér, A. (1981). Individual development and environment: A longitudinal study in Sweden. In S. A. Mednick & A. E. Baert (Eds.), *Prospective longitudinal research: An empirical basis for the primary prevention of psychosocial disorders*. Oxford: University Press.

Magnusson, D., Dunér, A., & Beckne, R. (1965). Planning. *Reports from IDA*. Department of Psychology, University of Stockholm, No. I. (In Swedish)

Magnusson, D., Dunér, A., & Olofsson, B. (1968). Criminal behavior: Models and research planning. *Reports from IDA*. Department of Psychology, University of Stockholm, No. 9. (In Swedish)

Magnusson, D., Dunér, A., & Zetterblom, G. (1968). The process of selecting a career. Models and research planning. *Reports from IDA*. Department of Psychology, University of Stockholm, No. 8. (In Swedish)

Magnusson, D., Dunér, A., & Zetterblom, G. (1975). *Adjustment: A longitudinal study*. New York: Wiley.

Magnusson, D., & Ekehammar, B. (1973). An analysis of situational dimensions: A replication. *Multivariate Behavioral Research, 8*, 331–339.

Magnusson, D., & Endler, N. S. (Eds.), (1977a). *Personality at the crossroads: Current issues in interactional psychology*. Hillsdale, N.J.: Lawrence Erlbaum Associates.

Magnusson, D., & Endler, N. S. (1977b). Interactional psychology: Present status and future prospects. In D. Magnusson & N. S. Endler (Eds.), *Personality at the crossroads*. Hillsdale, N.J.: Lawrence Erlbaum Associates.

Magnusson, D., Gerzén, M., & Nyman, B. (1968). The generality of behavioral data I. Generalization from observations on one occasion. *Multivariate Behavioral Research, 3*, 295–320.

Magnusson, D., & Heffler, B. (1969). The generality of behaviorial data III: Generalization potential as a function of the number of observation instances. *Multivariate Behavioral Research, 4*, 29–42.

Magnusson, D., & af Klinteberg, B. (1986). Hyperactivity and physiological arousal in terms of adrenaline excretion. Department of Psychology, University of Stockholm, (manuscript).

Magnusson, D., & Öhman, A. (Eds.) (1987). *Psychopathology: An interactional perspective*. New York: Academic Press.

Magnusson, D., & Olah, A. (1981). Situation-outcome contingencies: A study of anxiety provoking situations in a developmental perspective. *Reports from the Department of Psychology, University of Stockholm*, No. 579.

Magnusson, D., & Stattin, H. (1978). A cross-cultural comparison of anxiety responses in an interactional frame of reference. *International Journal of Psychology, 13*, 317–332.

Magnusson, D., & Stattin, H. (1981a). Methods for studying stressful situations. In W. H. Krohne & L. Laux (Eds.), *Achievement, stress and anxiety*. Washington, D.C.: Hemisphere.

Magnusson, D., & Stattin, H. (1981b). Situation-outcome contingencies: A conceptual and empirical analysis of threatening situations. *Reports from the Department of Psychology, University of Stockholm*, No. 571.

Magnusson, D., & Stattin, H. (1981c). Stability of cross-situational patterns of behavior. *Journal of Research in Personality, 15*, 481–496.

Magnusson, D., & Stattin, H. (1982). Biological age, environment and behavior in interaction: A methodological problem. *Reports from the Department of Psychology, University of Stockholm, No. 587*.

Magnusson, D., Stattin, H., & Allen, V. L. (1986a). Differential maturation among girls and its relation to social adjustment: A longitudinal perspective. In D. L. Featherman & R. M. Lerner (Eds.), *Life-span development*, Vol. 7. New York: Academic Press.

Magnusson, D., Stattin, H., & Allen, V. L. (1986b). Biological maturation and social development: A longitudinal study of some adjustment processes from mid-adolescence to adulthood. *Journal of Youth and Adolescence*, 14, 267–283.

Magnusson, D., & Törestad, B. (1982). Frequency and intensity of anxiety reactions. *Reports from the Department of Psychology, University of Stockholm*, No. 559.

Maier, S. F., & Landenslager, M. (1985). Stress and health: Exploring the links. *Psychology Today*, August, 44–49.

Mayr, E. (1976). *Evolution and the diversity of life*. Cambridge, Mass.: Harvard University Press.

Mead, G. H. (1934). *Mind, self and society*. Chicago: University of Chicago Press.

Mednick, S. A., Moffitt, T. E., Pollock, V., Talovic, S., Gabrielli, W. F., & Van Dusen, K. T. (1983). The inheritance of human deviance. In D. Magnusson & V. L. Allen (Eds.), *Human Development: An interactional perspective*. New York: Academic Press.

Meehl, P. E. (1978). Theoretical risks and tabular asterisks: Sir Karl, Sir Ronald, and the slow progress of soft psychology. *Journal of Consulting and Clinical Psychology*, 46, 806–834.

Meyer-Probst, B., Rösler, H. D., & Teichmann, H. (1983). Biological and psychosocial risk factors and development during childhood. In D. Magnusson & V. L. Allen (Eds.), *Human development: An interactional perspective*. Hillsdale, N.J.: Lawrence Erlbaum Associates.

Milich, R., Loney, J., & Landau, S. (1982). The independent dimensions of hyperactivity and aggression: A replication and further validation. *Journal of Abnormal Psychology*, 91, 183–198.

Miller, J. G. (1978). *Living systems*. New York: Mc Graw Hill.

Mineka, S., & Kihlstrom, J. F. (1978). Unpredictable and uncontrollable events: A new perspective on experimental neurosis. *Journal of Abnormal Psychology*, 87, 256–271.

Mischel, W. (1968). *Personality and assessment*. New York: Wiley.

Mischel, W., & Peake, P. K. (1982). Beyond deja vu in the search for cross-situational consistency. *Psychological Review*, 89, 730–755.

Mitchell, S., & Rose, P. (1981). Boyhood behavior problems as precursors of criminality: A fifteen-year follow-up study. *Journal of Child Psychology*, 22, 19–33.

Morris, H. H., Escoll, P. J., & Wexter, M. S. W. (1956). Aggressive behavior disorders in children. A follow-up study. *American Journal of Psychiatry*, 112, 991–997.

Moss, H. A., & Susman, E. J. (1980). Longitudinal study of personality development. In O. G. Brim, Jr. & J. Kagan (Eds.), *Constancy and change in human development*. Cambridge, Mass.: Harvard University Press.

Mulaik, S. A. (1964). Are personality factors raters' conceptual factors? *Journal of Consulting Psychology*, 28, 506–511.

Mulligan, G., Douglas, J. W. B., Hammond, W. A., & Tizard, J. (1963). Delinquency and symptoms of maladjustment: The findings from a longitudinal study. *Proceedings from the Royal Society of Medicine*, 56, 1083–1086.

Mumford, M. D., & Owens, W. A. (1984). Individuality in a developmental context: Some empirical and theoretical considerations. *Human Development*, 27, 84–108.

Murray, H. A. (1938). *Explorations in personality*. New York: Oxford University Press.

Mussen, P. H., & Jones, M. C. (1957). Self-conceptions, motivations and interpersonal attitudes of late- and early maturing boys. *Child Development*, 28, 243–256.

Mussen, P. H., & Young, H. B. (1964). Relationships between rate of physical maturity and personality among boys of Italian descent. *Vita Humana*, 7, 186–200.

Nesselroade, J. R., & Baltes, P. B. (Eds.), (1979). *Longitudinal methodology in the study of behavior and development*. New York: Academic Press.

Nesselroade, J. R., & Ford, D. H. (1985). P-technique comes of age: Multivariate, replicated, single-subject designs for research on older subjects. *Research on Aging*, 7, 46–80.

Nunnally, J. C. (1967). *Psychometric theory*. New York: McGraw-Hill.

Nygård, R. (1984). Toward an interactional psychology: Models from achievement motivation research. *Journal of Personality*, *49*, 363–387.

Nylander, I., & Rydelius, P. A. (1973). The relapse of drunkenness in non-asocial teen-age boys. *Acta Psychiatrica Scandinavia*, *49*, 435–443.

Nystedt, L. (1981). A model for studying the interaction between the objective situation and a person's construction of the situation. In D. Magnusson (Ed.), *Toward a psychology of situations: An interactional perspective*. Hillsdale: New Jersey: Lawrence Erlbaum Associates.

Offord, D. R., Sullivan, K., Allen, N., & Abrams, N. (1979). Delinquency and hyperactivity. *The Journal of Neurosis and Mental Disease*, *167*.

Öhman, A. (1986). Face the beast and fear the face: Animal and social fears as prototypes for evolutionary analyses of emotion. *Psychophysiology*, *23*, 123–145.

Öhman, A. (In press). The psychophysiology of emotion: An evolutionary-cognitive perspective. In P. K. Ackles, J. R. Jennings & M. G. H. Coles (Eds.), *Advances in psychophysiology*. Vol. 2 Greenwich, C.T.: JAJ Press.

Öhman, A., & Magnusson, D. (1987). An interactional paradigm for research on psychopathology. In D. Magnusson & A. Öhman (Eds.), *Psychopathology: An interactional perspective*. New York: Academic Press.

Olweus, D. (1979). Stability of aggressive reaction patterns in males: A review. *Psychological Bulletin*, *86*, 852–875.

Olweus, D. (1985). Aggression and hormones. Behavioral relationships with testosterone and adrenaline. In D. Olweus, J. Block, & M. Radke-Yarrow (Eds.), *The development of antisocial and prosocial behavior: Research, theories, and issues*. New York: Academic Press.

Overton, W. F., & Reese, H. W. (1973). Models of development: Methodological implications. In J. R. Nesselroade & H. W. Reese (Eds.), *Life-span developmental psychology: Methodological issues*. New York: Academic Press.

Parke, R. D. (1978). Parent-infant interaction: Progress paradigms and problems. In G. P. Sackett (Ed.), *Observing Behavior (Vol. 1): Theory and applications in mental retardation*. Baltimore: University Park Press.

Passini, F. T., & Norman, W. T. (1966). A universal conception of personality structure. *Journal of Personality and Social Psychology*, *4*, 44–49.

Pattersson, G. R. (1986). Performance models for anti-social boys. *American Psychologist*, *41*, 432–444.

Patterson, G. R., & Moore, D. R. (1978). Interactive patterns as units. In S. J. Suomi, M. E. Lamb & R. G. Stevenson (Eds.), *The Study of social interaction: Methodological issues*. Wisconsin: University of Wisconsin Press.

Pepper, S. C. (1942). *World hypotheses: A study in evidence*. Berkeley: University of California Press.

Pervin, L. (1968). Performance and satisfaction as a function of individual environment fit. *Psychological Bulletin*, *69*, 56–68.

Pervin, L. (1978). Theoretical approaches to the analysis of individual-environment interaction. In L. A. Pervin & M. Lewis (Eds.), *Perspectives in interactional psychology*. New York: Plenum.

Pervin, L. (1983). The stasis and flow of behavior. Toward a theory of goals. In M. M. Page (Ed.), *Personality: Current Theory and Research*. Lincoln: University of Nebraska Press.

Peskin, H. (1967). Puberal onset and ego functioning. *Journal of Abnormal Psychology*, *75*, 1–15.

Peterson, D. R. (1968). *The clinical study of social behavior*. New York: Appleton-Century-Crofts.

Peterson, D. R. (1979). Assessing interpersonal relationships by means of interaction research. *Behavioral Assessment*, *I*, 221–276.

Petersen, A. C., & Taylor, B. (1980). The biological approach to adolescence: Biological change and psychological adaptation. In J. Adelson (Ed.), *Handbook of adolescent psychology*. New York: Wiley.

Piaget, J. (1928). *Judgment and reasoning in the child*. New York: Harcourt, Brace.

Plomin, R. (1986). Behavioral genetic methods. *Journal of Personality*, 54, 226–261.

Plomin, R., & Daniels, D. (1984). The interaction between temperament and environment: Methodological considerations. *Merrill-Palmer Quarterly*, 30, 149–162.

Plomin, R., De Fries, J. C., & Loehlin, J. C. (1977). Genotype—environment interaction and correlation in the analysis of human behavior. *Psychological Bulletin*, 84, 309–322.

Pollack, R. H. (1983). Regression revisited: Perceptuo-cognitive performance in the aged. In S. Wapner & B. Kaplan (Eds.), *Toward a holistic developmental psychology*. Hillsdale, N.J.: Lawrence Erlbaum Associates.

Prechtl, H. F. R. (1976). Leichte frühkindliche Hirnschädigung und das Kompensationsvermögen des Nervsystems. *Bulletin der Schweizerischen Akademie der Medizinischen Wissenschaften*, 32, 99–113.

Pulkkinen, L. (1982). Self-control and continuity for childhood adolescence. Baltes, In P. B. Baltes & O. G. Brim Jr., (Eds.), *Life-span development and behavior*. Vol. 4. Academic Press, New York.

Radke-Yarrow, M., & Kuczynski, L. (1983). Conceptions of environment in childrearing interactions. In Magnusson, D. & Allen, V. L. (Eds.), *Human development: An interactional perspective*. New York: Academic Press.

Raleigh, M. J., Mc Guire, M. T., Brammer, G. L., & Yuwiler. (1984). Social and environmental influences on blood serotonin concentrations in monkeys. *Archives of General Psychiatry*, 41, 405–410.

Rasmuson, M. (1983). The role of genes as determinants of behavior. In D. Magnusson & V. L. Allen (Eds.), *Human development. An interactional perspective*. New York: Academic Press.

Rathjan, D. P., & Foreyt, J. P. (1980). *Social competence: Interventions for children and adults. Oxford: Pergamon Press.*

Raush, H. L. (1977). Paradox levels and junctions in person-situation systems. In D. Magnusson & N. S. Endler (Eds.), *Personality at the cross-roads: Current issues in interactional psychology*. Hillsdale, N.J.: Lawrence Erlbaum Associates.

Rausch, H. L. (1965). Interaction sequences. *Journal of Personality and Social Psychology*, 2, 487–499.

Raush, H. L., Dittmann, A. T., & Taylor, T. J. (1959). Person, setting and change in social interaction. *Human Relations*, 12, 361–379.

Raush, H. L., Farbman, I., & Llewellyn, L. G. (1960). Person, setting and change in social interaction: II. A normal-control study. *Human Relations*, 13, 305–333.

Riegel, K. E. (1978). *Psychology, mon amour*. Boston: Houghton Mifflin.

Riley, D., & Eckenrode. (1986). Social ties: Subgroup differences in costs and benefits. *Journal of Personality and Social Psychology*, 51, 770–778.

Robins, L. N. (1966). *Deviant children grow up*. Baltimore: Williams & Wilkens.

Robins, L. N. (1978). Aetiological implications in studies of childhood histories, relating to antisocial personality. In R. D. Hare & D. Schalling (Eds.), *Psychopathic behavior: Approaches to research*. New York: Wiley.

Roessler, R., Burch, N. R., & Mefferd, Jr., R. B. (1967). Personality correlates of catecholamine excretion under stress. *Journal of Psychosomatic Research*, 11, 181–185.

Roff, J. D., & Wirt, R. D. (1984). Childhood aggression and social adjustment as antecedants of delinquency. *Journal of Abnormal Child Psychology*, Vol. 12, 111–126.

Rotter, J. B. (1954). *Social learning and clinical psychology*. Englewood Cliffs, N.J.: Prentice-Hall.

Rotter, J. B. (1955). The role of the psychological situation in determining the direction of human behavior. In M. R. Jones (Ed.). *Nebraska Symposium on Motivation*. Lincoln: University of Nebraska Press.

Rushton, J. P. (1984). Sociobiology: Toward a theory of individual and group differences in personality and social behavior. In J. R. Royce & L. P. Mos (Eds.), *Annals of theoretical psychology* (Vol. 2). New York: Plenum.

Rushton, J. P., Fulker, D. W., Neale, M. C., Nias, D. K. B., & Eysenck, H. J. (1986). Altruism and aggression: Individual differences are substantially heritable. *Journal of Personality & Social Psychology*, 50, 1192–1198.

Russell, R. W. (1970). "Psychology": Noun or adjective. *American Psychologist*, *25*, 211–218.

Rutter, M. (1972). Relationship between child and adult psychiatric disorders. *Acta Psychiatrica Scandinavica*, *48*, 3–21.

Rutter, M. (1979). Protective factors in children's responses to stress and disadvantage. In M. W. Kent & J. E. Rolf (Eds.), *Primary preventions of psychopathology: Social competence in children* (Vol. 3). Hanover, NH: University Press of New England.

Rutter, M. (1982). Epidemiological-longitudinal approaches to the study of development. In W. A. Collins (Ed.), *The concept of development. The Minnesota Symposia on Child Psychology*, Vol. 15. Hillsdale, N.J.: Lawrence Erlbaum Associates.

Rutter, M. (1983). Statistical and personal interactions: Facets and perspectives. In D. Magnusson & V. L. Allen (Eds.), *Human development: An interactional perspective*. New York: Academic Press.

Rutter, M., & Giller, H. (1983). *Juvenile delinquency: Trends and perspectives*. Harmondsworth, Middlesex: Penguin.

Rutter, M., & Graham, P. (1968). The reliability and validity of the psychiatric assessment of the child: Interview with child. *British Journal of Psychiatry*, *114*, 563–579.

Rutter, M., Cox, A., Tupling, C., Berger, M., & Yule, W. (1975). Attainment and adjustment in two geographical areas. The prevalence of psychiatric disorder. *British Journal of Psychiatry*, *126*, 493–509.

Rutter, M., Tizard, J., & Whitmore, K. (1970). *Education, health and behavior*. London: Longman.

Rychlak, J. F. (1981). The case for a modest revolution in modern psychological science. In R. A. Kasschan & Ch. N. Cofer (Eds.), *Psychology's second century: Enduring issues*. New York: Praeger.

Rydelius, P. A. (1978). Barnpsykiatriskt omhändertagande av unga fyllerister. (Young drunkards in child psychiatric care). *Läkartidningen*, *75*, 1607–1611 (in Swedish).

Sameroff, A. J. (1975). Early influences on development: Fact or fancy? *Merrill-Palmer Quarterly*, *21*, 267–294.

Sameroff, A. J. (1982). Development and the dialectic: The need for a systems approach. In W. A. Collins (Ed.), *The concept of development*. Hillsdale, N. J.: Lawrence Erlbaum Associates.

Sameroff, A. J. (1983). Developmental systems: Contexts and evolution. In P. H. Mussen (Ed.), *Handbook of Child Psychology*, Vol. 1. New York: Wiley.

Satterfield, J. H., Cantwell, D. P., & Satterfield, B. T. (1974). Pathophysiology of the hyperactive child syndrome. *Archives of General Psychiatry*, *31*, 839–844.

Satterfield, J. H., & Dawson, M. E. (1971). Electrodermal correlates of hyperactivity in children. *Psychophysiology*, 8, 191–198.

Satterfield, J. H., Hoppe, C. M., & Schell, A. M. (1982). A prospective study of delinquency in 110 adolescent boys with attention deficit disorder and 88 normal adolescent boys. *American Journal of Psychiatry*, 139:6, 795–798.

Scarr, S. (1981). Comments on psychology: Behavior genetics and social policy from an anti-reductionist. In R. A. Kasschan & Ch. N. Cofer (Eds.), *Psychology's second century: Enduring issues*. New York: Praeger.

Scarr, S., & McCartney, K. (1983). How people make their own environments: A theory of genotype → environment effects. *Child Development*, *54*, 424–435.

Scarr-Salapatek, S. (1971). Race, social class, and IQ. *Science*, *174*, 1285–1295.

Schaie, K. W. (1965). A general model for the study of developmental problems. *Psychological Bulletin*, *64*, 92–107.

Schaie, K. W. (1972). Limitations in the generalizability of growth curves of intelligence. *Human Development*, *15*, 141–152.

Schaie, K. W., & Baltes, P. B. (1975). On sequential strategies in developmental research: Description or explanation? *Human Development, 18*, 384–390.

Schachar, R., Rutter, M., & Smith, A. (1981). The characteristics of situationally and pervasively hyperactive children: Implications for syndrome definition. *Journal of Child Psychology and Psychiatry, 22*, 375–392.

Schneirla, T. C. (1957). The concept of development in comparative psychology. In D. B. Harris (Ed.), *The concept of development*. Minneapolis: University of Minnesota Press.

Schneirla, T. C. (1972). Levels in the psychological capacities of animals. In L. R. Aronson, E. Tobach, J. S. Rosenblatt & D. S. Lehrman (Eds.), *Selected writings of T. C. Schneirla*. San Francisco: Freeman.

Schweder, R. A. (1975). How relevant is an individual difference theory of personality? *Journal of Personality, 43*, 455–484.

Scriven, M. (1959). Explanation and prediction in evolutionary theory. *Science 130*, 477–482.

Sears, R. R. (1951). A theoretical framework for personality and social behavior. *American Psychologist, 6*, 476–483.

Seligman, M. E. P. (1975). *Helplessness: On depression, development and death*. San Francisco: W. H. Freeman & Company.

Sells, S. B. (1963). An interactionist looks at the environment. *American Psychologist, 18*, 696–702.

Sells, S. B. (1966). Ecology and the science of psychology. *Multivariate Behavioral Research, 1*, 131–144.

Shantz, C. U. (1983). Social cognition. In P. H. Mussen (Ed.), *Handbook of Child Psychology*. Vol. 3. New York: Wiley.

Simmons, R. G., Blyth, D. A., & Mc Kinney, K. L. (1983). The social and psychological effects of puberty on white males. In J. Brooks-Gunn & A. C. Petersen (Eds.), *Girls at puberty: Biological and psychological perspectives*. New York: Plenum.

Simmons, R. G., Blyth, D. A., Van Cleave, E. F., & Bush, D. M. (1979). Entry into early adolescence: The impact of school structure, puberty and early dating on self-esteem. *American Sociological Review, 44*, 948–967.

Sjöberg, L. (1982). Attitude-behavior correlation, social desirability and perceived diagnostic value. *British Journal of Social Psychology, 21*, 283–292.

Sjöbring, H. (1958). *Structure and development: A personality theory*. Lund: Gleerups. (In Swedish)

Skinner, B. F. (1971). *Beyond freedom and dignity*. New York: Knopf.

Sparks, R., Genn, H. G., & Dodd, D. J. (1977). *Surveying victims. Measurement of criminal victimization, perceptions of crime and attitudes to crimnal justice*. London: Wiley.

Sperry, R. W. (1982). Some effects of disconnecting the cerebral hemispheres. *Science, 217*, 1223–1226.

Spielberger, C. D. (1977). State-trait anxiety and interactional psychology. In D. Magnusson & N. S. Endler (Eds.), *Personality at the crossroads: Current issues in interactional psychology*. Hillsdale, N.J.: Lawrence Erlbaum Associates.

Sroufe, L. A. (1979a). The coherence of individual development. Early care, attachment and subsequent developmental issues. *American Psychologist, 34*, 834–841.

Sroufe, L. A. (1979b). Socioemotional development. In J. Osofsky (Ed.), *Handbook of infant development*. New York: Wiley.

Sroufe, L. A., & Rutter, M. (1984). The domain of developmental psychopathology. *Child Development, 55*, 17–29.

Staats, A. W. (1963). *Complex human behavior*. New York: Holt, Rinehart & Winston.

Staats, A. W. (1981). Paradigmatic behaviorism, unified theory, unified theory construction, methods and the zeitgeist of separatism. *American Psychologist, 36*, 239–256.

Stagner, R. (1977). On the reality and relevance of traits. *The Journal of General Psychology, 96*, 185–207.

Stattin, H. (1983). *The psychological situation in an interactional perspective of personality: A theoretical background and some empirical studies.* Doctoral Dissertation, Department of Psychology, University of Stockholm.

Stattin, H., & Magnusson, D. (1981). Situation-outcome contingencies of threatening experiences: Age and sex differences. *Reports from the Department of Psychology, University of Stockholm.* No. 580.

Stattin, H., & Magnusson, D. (1984). The role of early aggressive behavior for the frequency, the seriousness and the types of later criminal offenses. *Reports from the Department of Psychology, University of Stockholm.* No. 618.

Stattin, H., & Magnusson, D. (in press). *Early maturing girls.* Vol. 2 in D. Magnusson (Ed.), Paths through life. Hillsdale, N.J.: Erlbaum.

Stattin, H., Magnusson, D., & Reichel, H. (1986). Criminality from Childhood to Adulthood. *Reports from IDA, Department of Psychology, University of Stockholm.* No. 63.

Stein, M. (1985). Bereavement, depression, stress and immunity. In R. Guillemin et al. (Eds.), *Neural modulation of immunity,* New York: Raven Press.

Stein, M., & Schleifer, S. J. (1985). Frontier of stress research: Stress and immunity. In M. Zales (Ed.), *Stress in health and disease.* New York: Bruner/Mazel.

Stewart, M. A., Cummings, C., Singer, S., & De Blois, C. S. (1981). The overlap between hyperactive and unsocialized aggressive children. *Journal of Child Psychology and Psychiatry, 22,* 35–45.

Suomi, S. J. (1979). Peers, play and primary prevention in primates. In M. W. Kent & E. Roy (Eds.), *Primary prevention of psychopathology: Social competence in children* (Vol. 3.). Hanover, N.H.: University Press of New England.

Takala, M. (1984). A socio-ecological approach to personality and the problem of situations. In H. Bovarius, G. van Heck & N. Smid (Eds.), *Personality psychology in Europe.* Amsterdam: Swets & Zeitlinger.

Tanner, J. M. (1978). *Foetus into man: Physical growth from conception to maturity.* London: Open Books.

Thomas, W. I. (1927). The behavior pattern and the situation. *Publications of the American Sociological Society: Papers and Proceedings.* (Vol. 22).

Thomas, W. I. (1928). *The child in America.* New York: Knopf.

Thomas, A., Birch, H. G., Chess, S., & Robbins, L. C. (1961). Individuality in responses of children to similar environmental situations. *American Journal of Psychiatry, 2,* 236–245.

Thomas, A., & Chess, S. (1977). *Temperament and development.* New York: Bruner/Mazel.

Thomas, A., & Chess, S. (1980). *The dynamics of psychological development.* New York: Bruner/Mazel.

Tolman, E. C. (1949). *Purposive behavior and men.* Berkeley, University of California Press.

Tolman, E. C. (1951). A psychological model. In T. Parsons & E. A. Shils (Eds.), *Toward a general theory of action.* Cambridge, Mass.: Harvard University Press.

Toulmin, S. (1981). Toward reintegration: An agenda for psychology's second century. In R. A. Kasschau & Ch. N. Cofer (Eds.), *Psychology's second century: Enduring issues.* New York: Praeger.

Törestad, B., Magnusson, D., & Olah, A. (1984). Individual control, intensity of reactions and frequency of occurrence. *Reports from the Department of Psychology, University of Stockholm,* Report no. 630.

Törestad, B., Magnusson, D., & Olah, A. (1985). Coping, control and stress experiences: An interactional perspective. *Reports from the Department of Psychology, University of Stockholm,* Report no. 643.

Ulvund, S. E. (1980). Cognition and motivation in early infancy: An interactionistic approach. *Human Development, 23,* 17–32.

Underwood, B. J. (1975). Individual differences as a crucible in theory construction. *American Psychologist, 30,* 128–134.

Urban, H. B. (1978). The concept of development from a systems perspective. In P. Baltes (Ed.), *Life span development and behavior*. Vol. 1 New York: Academic Press.

Uzgiris, I. C. (1977). Plasticity and structure. The role of experience in infancy. In I. C. Uzgiris & F. Weizmann (Eds.), *The structuring of experience*, New York: Plenum.

Vale, J. R., & Vale, G. R. (1969). Individual differences and general laws in psychology: A reconciliation. *American Psychologist*, *24*, 1093–1108.

Vernon, P. E. (1964). *Personality assessment: A critical survey*. New York: Wiley.

Vikan, A. (1985). Psychiatric epidemiology in a sample of 1510 ten-year-old children-I. Prevalence. *Journal of Child Psychology and Psychiatry*, *26*, 55–75.

Wachs, T. D. (1977). The optimal stimulation hypothesis and early development: Anybody got a match? In I. C. Uzgiris & F. Weizmann (Eds.), *The structuring of experience*. New York: Plenum.

Wadsworth, M. E. J. (1979). *Roots of delinquency: Infancy, adolescence and issues*. London: Robertson.

Waid, W. M. (1984). *Sociophysiology*. New York: Springer Verlag.

Wapner, S., & Kaplan, B. (1983). (Eds.), *Toward a holistic developmental psychology*. Hillsdale, N.J.: Lawrence Erlbaum Associates.

Watson, J. B. (1913). Psychology as the behaviorist views it. *Psychological Review*, *20*, 158–177.

Watson, J. B. (1930). *Behaviorism* (2nd ed.). New York: Norton.

Wechsler, H. (1979). Patterns of alcohol consumption among the young: High school, college and general population studies. In H. Blane & M. Chafez (Eds.), *Youth, alcohol and social policy*. New York: Plenum Press.

Weiner, H. (1977). *Psychology and human disease*. New York: Elsevier.

Weiss, P. A. (1969). The living system: Determinism stratified. In A. Koestler & J. R. Smythies (Eds.), *Beyond reductionism: New perspectives in the life sciences*. New York: Mac Millan.

Weisz, J. R. (1983). Can I control it? The pursuit of veridical answers across the life span. In P. B. Baltes & O. G. Brim (Eds.), *Life-span development and behavior*. Vol 5. New York: Academic Press.

Wenar, Ch. (1982). Developmental psychopathology: Its nature and models. *Journal of Clinical Child Psychology*, *11*, 192–201.

Werner, E. E., & Smith, R. S. (1982). *Vulnerable but invincible. A study of resilient children*. New York: Mc Graw-Hill.

West, D. J. (1982). *Delinquency: Its roots, career and prospects*. London: Heinemann Educational Books.

West, D. J., & Farrington, D. P. (1973). *Who becomes delinquent?* Second report of the Cambridge study in delinquent development. London: Heinemann.

Westin-Lindgren, G. (1979). *Physical and mental development in Swedish urban school children*. Lund: CWK/Gleerup.

Wigner, E. P. (1969). Epistemology of quantum mechanics: It's appraisals and demands. In M. Grene (Ed.), *The anatomy of knowledge*. London: Routledge & Kegan Paul.

Wilson, E. O. (1975). *Sociobiology: The new synthesis*. Cambridge, MA: Harvard University Press.

Wilson, J. Q., & Herrnstein, R. J. (1985). *Crime and human nature*. New York: Simon & Schuster.

de Wit, J. (1986). Adolescence: Socialization issues and the impact of values. In L. Y. Ching, Ch. H. Keng, & L. Ch. S. Men (Eds.), *Preparation for adulthood*. Department of Pedagogy and Educational Psychology, University of Malaya.

Wohlwill, J. F. (1970). The age variable in psychological research. *Psychology Review*, *77*, 49–64.

Wohlwill, J. F. (1973). *The study of behavioral development*. London: Academic Press.

Wohlwill, J. F. (1980). Stability of cognitive development in childhood. In O. G. Brim & J. Kagan (Eds.), *Constancy and change in human development*. Cambridge, Mass.: Harvard University Press.

Wohlwill, J. F. (1981). Physical and social environment as factors in development. In D. Magnusson (Ed.), *Toward a psychology of situations: An interactional perspective*. Hillsdale, New Jersey: Lawrence Erlbaum Associates.

Wolff, P. H. (1981). Normal variation in human maturation. In K. J. Conolly & H. F. R. Prechtl (Eds.), *Maturation and development: Biological and psychological maturation*. London: Heinemann Medical Books.

Woodman, D., Hinton, J., & O'Neill, M. (1977). Relationship between violence and catecholamines. *Perceptual and Motors Skills, 45*, 702.

Wundt, W. (1948). Principles of physiological psychology. In W. Dennis (Ed.), *Readings in the history of psychology*. New York: Appleton-Century-Crofts.

Yarrow, M. R., Campbell, J. O., & Burton, R. V. (1970). Recollections of childhood: A study of the retrospective method. *Monographs of the Society for Research in Child Development, 35*, No. 5.

Yarrow, L. J., Rubenstein, J. L., & Pedersen, F. A. (1975). *Infant and environment: Early cognitive and motivational development*. New York: Wiley (Halsted Press).

Zajonc, R. B. (1984). On the primacy of affect. *American Psychologist, 39*, 117–123.

Zettergren, P. (1979). Social rejection and isolation in a longitudinal perspective. *Reports from the Department of Psychology, University of Stockholm*, Report No. 31 (In Swedish).

Zettergren, P. (1980). Social situation and development—a study of rejected and isolated children. *Reports from the Department of Psychology, University of Stockholm*, Report No. 36 (In Swedish).

Zettergren, P., & Dunér, A. (1979). Socially rejected and isolated in the school class. *Reports from the Department of Psychology, University of Stockholm*, Report No. 32 (In Swedish).

Zubin, J., & Spring, B. (1977). Vulnerability: A review of schizophrenia. *Journal of Abnormal Psychology, 186*, 103–126.

Zubin, J., & Steinhauer, S. (1981). How to break the logjam in schizophrenia: A look beyond genetics. *The Journal of Nervous Mental Disease, 169*, 477–494.

Zucker, R. (1979). Developmental aspects of drinking through the young adult years. In H. Blane & M. Chafez (Eds.), *Youth, alcohol and social policy*. New York: Plenum Press.

Zuckerman, M. (1979). *Sensation seeking*. Hillsdale, N.J.: Lawrence Erlbaum Associates.

AUTHOR INDEX

W

Y

Z

SUBJECT INDEX